PYROLYSIS–GAS CHROMATOGRAPHY

Mass Spectrometry of Polymeric Materials

Other Related Titles from World Scientific

Fast Liquid Chromatography–Mass Spectrometry Methods in Food and Environmental Analysis
edited by Oscar Núñez, Héctor Gallart-Ayala, Claudia PB Martins and Paolo Lucci
ISBN: 978-1-78326-493-3

Fundamentals of Electrothermal Atomic Absorption Spectrometry: A Look Inside the Fundamental Processes in ETAAS
by A-Javier Aller
ISBN: 978-981-3229-76-1

High-Performance Liquid Chromatography and Mass Spectrometry of Porphyrins, Chlorophylls and Bilins
by Chang Kee Lim
ISBN: 978-981-02-3068-5

High Performance Liquid Chromatography Fingerprinting Technology of the Commonly-Used Traditional Chinese Medicine Herbs
by Baochang Cai, Seng Poon Ong and Xunhong Liu
ISBN: 978-981-4291-09-5

Problems of Instrumental Analytical Chemistry: A Hands-On Guide
by JM Andrade-Garda, A Carlosena-Zubieta, MP Gómez-Carracedo, MA Maestro-Saavedra, MC Prieto-Blanco and RM Soto-Ferreiro
ISBN: 978-1-78634-179-2
ISBN: 978-1-78634-180-8 (pbk)

Recent Developments in Plasmon-Supported Raman Spectroscopy: 45 Years of Enhanced Raman Signals
edited by Katrin Kneipp, Yukihiro Ozaki and Zhong-Qun Tian
ISBN: 978-1-78634-423-6

Software-Assisted Method Development in High Performance Liquid Chromatography
edited by Szabolcs Fekete and Imre Molnár
ISBN: 978-1-78634-545-5

PYROLYSIS–GAS CHROMATOGRAPHY

Mass Spectrometry of Polymeric Materials

Peter Kusch

Bonn-Rhein-Sieg University of Applied Sciences, Germany

NEW JERSEY • LONDON • SINGAPORE • BEIJING • SHANGHAI • HONG KONG • TAIPEI • CHENNAI • TOKYO

Published by

World Scientific Publishing Europe Ltd.

57 Shelton Street, Covent Garden, London WC2H 9HE

Head office: 5 Toh Tuck Link, Singapore 596224

USA office: 27 Warren Street, Suite 401-402, Hackensack, NJ 07601

Library of Congress Cataloging-in-Publication Data
Names: Kusch, Peter, author.
Title: Pyrolysis-gas chromatography : mass spectrometry of polymeric materials /
by Peter Kusch, Bonn-Rhein-Sieg University of Applied Sciences, Germany.
Description: New Jersey : World Scientific, [2018] | Includes bibliographical references and index.
Identifiers: LCCN 2018023726 | ISBN 9781786345752 (hc : alk. paper)
Subjects: LCSH: Polymers--Analysis. | Gas chromatography. | Pyrolysis.
Classification: LCC QD139.P6 K87 2018 | DDC 547/.7--dc23
LC record available at https://lccn.loc.gov/2018023726

British Library Cataloguing-in-Publication Data
A catalogue record for this book is available from the British Library.

For any available supplementary material, please visit
https://www.worldscientific.com/worldscibooks/10.1142/Q0171#t=suppl

Desk Editors: Anthony Alexander/Jennifer Brough/Shi Ying Koe

Typeset by Stallion Press
Email: enquiries@stallionpress.com

Printed in Singapore

Preface

Analytical pyrolysis (Py) is defined as the characterization in an inert atmosphere of a high-molecular material or a chemical process by a chemical degradation reaction(s) induced by thermal energy. Thermal degradation under controlled conditions is often used as a part of an analytical procedure, either to render a sample into a suitable form for subsequent analysis by gas chromatography (GC), mass spectrometry (MS), gas chromatography coupled with the mass spectrometry (GC/MS), with the Fourier-transform infrared spectroscopy (GC/FTIR) or by direct monitoring as an analytical technique in its own right. Analytical pyrolysis deals with structural identification and quantitation of pyrolysis products, with the ultimate aim of establishing the identity of the original material and the mechanisms of its thermal decomposition. The methodology of analytical Py–GC/MS has been known for several decades but is not used in many analytical laboratories of research and process control in the industry. The reason for this is the relative difficulty of interpreting the identified pyrolysis products, as well as the variety of them. Py–GC/MS can be applied to research and development of new materials, quality control, characterization and competitor product evaluation, medicine, biology and biotechnology, geology, airspace, and environmental analysis to forensic purposes or conservation and restoration of cultural heritage. These applications cover analysis and identification of polymers/copolymers and additives in components of automobiles, tires, packaging materials, textile fibers, coatings, adhesives, half-finished products for electronics, paints or varnishes, lacquers, leather, paper or wood products, food, pharmaceuticals, surfactants, and fragrances.

My research at the Department of Applied Natural Sciences at the Bonn-Rhein-Sieg University of Applied Sciences in Rheinbach (Germany) focuses on the application of the analytical Py–GC/MS and headspace–solid-phase microextraction–GC/MS (HS–SPME–GC/MS) for characterization of polymeric materials and components from many branches of the manufacturing and building industries. In recent years, however, the focus has been on the characterization of polymeric materials and failure analysis, especially in the automotive industry. The high success rate for solving problems and the satisfaction of the clients have convinced me that this analytical technique is well suited for failure analysis in the automotive industry. The obtained analytical results were then used for troubleshooting and remedial action of the technological process. Some of the results obtained have been presented at international symposia and published in analytical journals. In my work, besides projects, I was also involved to conduct the practical courses in analytical pyrolysis for students of chemistry and polymer science from the Bonn-Rhein-Sieg University of Applied Sciences in Rheinbach and Aachen University of Applied Sciences in Jülich (Germany). Under the auspices of the German Chemical Society (Gesellschaft Deutscher Chemiker, Frankfurt), together with my colleagues Professor Gerd Knupp, Professor Margit Geissler, Dr. Johannes Steinhaus, and M.Sc. Lara Kehret, I was involved for over 10 years in the implementation of the courses "Application of Pyrolysis–Gas Chromatography/Mass Spectrometry for Characterization of Plastics" for participants from industry, research institutes, and academia from Germany, Austria, and Switzerland. The results, collected in over 20 years of experience in the field of analytical pyrolysis, became the basis for the experimental material of the book. The essay contains an introduction to the methodology of analytical pyrolysis. The common devices have been described. Emphasis is placed on the identification of pyrolysis products from different classes of synthetic polymers/copolymers and biopolymers. The results of identification of several classes of polymers/copolymers and biopolymers can be very helpful to the user of this technique. Practical applications of these hyphenated analytical techniques for the analysis of microplastics, for failure analysis in the automotive industry, and solving of technological problems in the industry were presented in the book. The demonstrated practical applications can encourage many analytical chemists and engineers to use this technique in their laboratories. I hope that the book will be a great help for analytical chemists and engineers who work in the field of plastic analytics.

Dr. Peter Kusch

About the Author

Dr. Peter Kusch studied chemistry at the Pedagogical University in Opole and doctorate in organic chemical technology at the Poznań University of Technology, Poland. From 1977 to 1988 he has worked as an analytical chemist and adjunct at the Institute of Heavy Organic Synthesis "Blachownia" (Kędzierzyn-Koźle, Poland). After moving with the family to Germany, he has worked for several years in the *Fischer* Labor- und Verfahrenstechnik GmbH company (Meckenheim/Bonn, Germany) as laboratory manager and specialist for the analytical pyrolysis and gas chromatography. Since 1998 he is a scientific co-worker at the Department of Applied Natural Sciences of the Bonn-Rhein-Sieg University of Applied Sciences in Rheinbach, Germany. He has been author/coauthor of over 90 scientific publications, 11 invited book chapters, and 11 patents in the area of chromatography, mass spectrometry, analytical pyrolysis, and chemical technology. Peter Kusch is reviewer for several international journals in the area of analytical chemistry. He is the Editorial Board Member of the journal *Polymer Testing* and the Member of the American Chemical Society (ACS).

Acknowledgments

I would like to thank Dr. Merlin Fox (World Scientific Publishing, London, UK) for his invitation to write this book, Ms. Jennifer Brough (World Scientific Publishing, London, UK) for her excellent care while writing and Mr. Suraj Kumar and Mr. Anthony Alexander (World Scientific Publishing, Chennai, India) for copyediting of the manuscript.

I thank Dr. Johannes Steinhaus, M.Sci. Lara Kehret and M.Eng. Esther van Dorp from the *Kompetenzplattform Polymere Materialien* at the Bonn-Rhein-Sieg University of Applied Sciences in Rheinbach, Germany, as well as Prof. Dr.-Ing. Dorothee Schroeder-Obst and Dr.-Ing. Volker Obst from the firm Dr. Obst Technische Werkstoffe GmbH (Rheinbach, Germany) for samples and for the excellent cooperation for several years.

I am grateful to Dr. Nkechi H. Okoye (Nnamdi Azikiwe University, Awka, Anambra State, Nigeria) for samples of tropical wood.

I acknowledge the permanent support and encouragement of my lovely family and the engagement of my son Matthaeus for critical reading and correction of the manuscript.

Contents

1
Introduction

The word *pyrolysis*, translated from the original Greek *pyros* = *fire* and *lyso* = *decomposition*, means a chemical transformation of a sample when heated at a temperature higher than ambient in an inert atmosphere in the absence of oxygen [1]. Pyrolysis can be divided into two types, applied pyrolysis and analytical pyrolysis [2]. Applied pyrolysis is concerned with the production of chemicals. When performed on a large scale, pyrolysis is involved in industrial processes as the manufacture of coke from coal and the conversion of biomass into biofuels. In contrast, analytical pyrolysis is a laboratory procedure in which small amounts of organic materials undergo thermal treatment. The pyrolysis itself is just a process that allows the transformation of the sample into other compounds. The pyrolytic process is carried out in a pyrolysis unit (pyrolyzer) interfaced with the analytical instrumentation.

1.1. Definition of the Analytical Pyrolysis

According to the International Union of Pure and Applied Chemistry (IUPAC) recommendation, analytical pyrolysis is defined as the characterization in an inert atmosphere of a material or a chemical process by a chemical degradation reaction(s) induced by thermal energy [3]. Thermal degradation under controlled conditions is often used as a part of an analytical procedure, either to render a sample into a suitable form for subsequent analysis by gas chromatography (GC), mass spectrometry (MS), gas chromatography coupled with the mass spectrometry (GC/MS), with the Fourier-transform infrared spectroscopy (GC/FTIR), or by direct monitoring as an analytical technique in its own right [3]. Analytical pyrolysis deals with the structural identification and quantitation of pyrolysis products with the ultimate aim of establishing the identity of the original material and the mechanisms of its thermal decomposition [4]. The pyrolysis temperatures of 550–1400°C are high enough to actually break molecular bonds in the molecules of the solid sample, thereby forming smaller, simpler volatile compounds. Depending on the amount of energy supplied, the bonds in each molecule break in a predictable manner [5]. Most of the degradation results from free radical reactions initiated by bond breaking and depends on the relative strengths of the bonds that hold the molecules together. A large molecule will break apart and rearrange in a characteristic way. If the energy transfer to the sample is controlled by temperature, heating rate, and time, the fragmentation pattern is reproducible and characteristic for the original polymer [6]. Another sample of the same composition heated at the same rate to the same temperature, for the same period of time, will produce the same fragments. By the identification and measurement of the fragments, the molecular composition of the original sample can often be reconstructed. Pyrolysis may not be thought of as a sample preparation technique, but more of a sample destructive technique since the actual molecular form of the sample is changed upon heating. The chronological development of the analytical pyrolysis is described in detail in the publication of Tsuge [7].

1.2. Application of the Analytical Pyrolysis

Pyrolysis is often used for the analysis of polymer and copolymer samples since they are too high of a molecular weight (MW) to analyze using GC techniques and they often degrade in a systematic fashion. The applications of the analytical pyrolysis–gas chromatography/mass spectrometry (Py–GC/MS) range from research and development of new materials, quality control, characterization and competitor product evaluation, medicine, biology and biotechnology, geology, airspace, environmental analysis to forensic purposes, or conservation and restoration of cultural heritage. These applications cover analysis and identification of polymers/copolymers and additives in components of automobiles, tires, packaging materials, textile fibers, coatings, adhesives, half-finished products for electronics, paints or varnishes, lacquers, leather, paper or wood products, food, pharmaceuticals, surfactants, and fragrances [8].

1.2.1. Analytical pyrolysis of synthetic organic polymers and copolymers

The word *polymer*, translated from original Greek, means *many parts*. Structural analysis and the study of the degradation properties of high-MW polymers, such as plastics, rubber, and resins, are important to understand and improve performance characteristics of a polymer in many industrial applications. Polymers cannot be analyzed in their normal state by traditional GC because of their high MW and lack of volatility.

Although synthetic polymers and copolymers are inherently difficult to analyze because of their high MW and lack of volatility, various analytical techniques are used to characterize polymers/copolymers. These techniques including physical testing (rheological testing), thermogravimetric analysis (TGA), electron microscopy, Fourier transform infrared (FTIR) spectroscopy, size-exclusion chromatography (SEC)/gel permeation chromatography (GPC), nuclear magnetic resonance (NMR), laser light scattering, ultraviolet–visible (UV/Vis) spectroscopy and MS. The listed nondestructive methods offer information about functional groups, structural elements, thermal stability, MW, and volatile components. They have limitations and often laborious and time-consuming sample preparation, including hydrolysis, dissolution, or derivatization, is needed before analysis. Analytical pyrolysis (Py) technique hyphenated to GC/MS has extended the range of possible tools for the characterization of synthetic polymers and copolymers. This technique has been used extensively over the last 30 years as a complementary analytical tool used to characterize the structure of synthetic organic polymers and copolymers, polymer blends, biopolymers, and natural resins. Pyrolysis–GC/MS is a destructive analytical technique. Typical fields of interest and application are [9–12]:

— polymer identification by comparison of pyrograms and mass spectra with known references;
— qualitative analysis and structural characterization of copolymers, sequence statistics of copolymers, differentiation between statistical and block polymers;
— determination of the (micro) structure of polymers (degree of branching and crosslinking, compositional analysis of copolymers and blends, co-monomer ratios, sequence distributions, analysis of end-groups);
— determination of the polymers steric structure (stereoregularity, tacticity, steric block length, and chemical inversions);
— investigation of thermal stability, degradation kinetics, and oxidative thermal decomposition of polymers and copolymers;
— determination of monomers, volatile organic compounds (VOCs), solvents, and additives in polymers;
— kinetic studies;
— quality control; and
— quantification.

2

Apparatus in Analytical Pyrolysis

Figure 2.1 illustrates a schematic diagram of a pyrolysis–gas chromatography/mass spectrometry (Py–GC/MS) measuring system with the direct connections of an oven pyrolyzer together with the temperature and pressure controller, a gas chromatograph (GC) equipped with a capillary separation column, and detection device using a quadrupole mass spectrometer (MS).

Pyrolysis method allows for the direct analysis of very small solid or liquid polymer/copolymer sample amounts (5–200 μg) and eliminates the

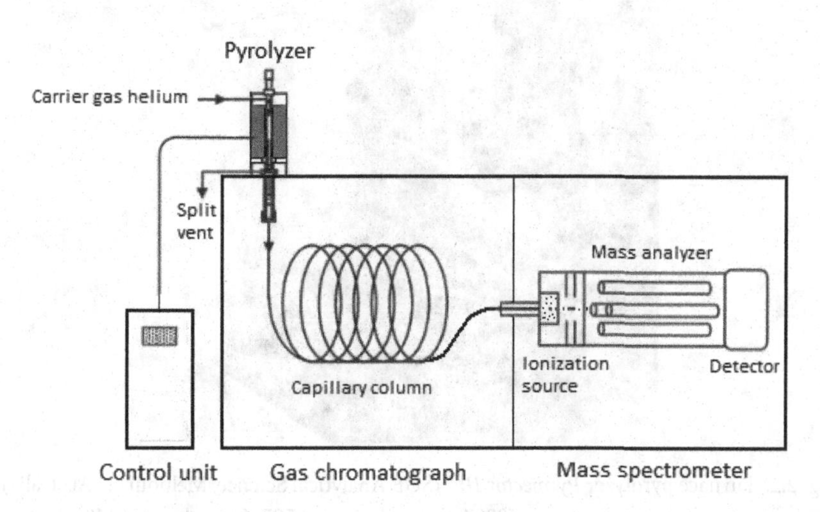

Fig. 2.1. Schematic diagram of a Py–GC/MS apparatus (based on Ref. [6]).

need for time-consuming sample preparation (pre-treatment) by performing analyses directly on the sample. When a polymer/copolymer sample is subjected to pyrolysis, primary bond-fission processes are initiated. These may proceed by means of several temperature-dependent and competing reactions, which make the final fragment distribution highly dependent upon the pyrolysis temperature. Therefore, the essential requirements for the apparatus in analytical pyrolysis are reproducibility of the final pyrolysis temperature, rapid temperature rise, and accurate temperature control. Depending upon the heating mechanism, pyrolysis systems have been classified into two groups: the continuous-mode pyrolyzer (furnace pyrolyzer) and pulse-mode pyrolyzer (flash pyrolyzer), such as the heated filament, Curie-point, and laser pyrolyzer. The pyrolysis unit is connected directly to the injector port of a GC (Figs. 2.2 and 2.3).

Fig. 2.2. Furnace pyrolyzer *Pyrojector II*™ (SGE Analytical Science, Melbourne, Australia) used in this work connected to a *7890A* GC and a series *5975C* quadrupole MS (Agilent Technologies Inc., Santa Clara, CA, USA).

Fig. 2.3. Furnace pyrolyzer *Pyrojector II*™ (SGE Analytical Science, Melbourne, Australia) used in this work connected to a *Trace 2000* GC (ThermoQuest CE Instruments, Milan, Italy) and a *Voyager* quadrupole MS (ThermoQuest/Finnugan MassLab Group, Manchester, UK).

Once the polymer/copolymer sample has been pyrolyzed, volatile fragments are swept from the heated pyrolysis unit by the carrier gas (helium) into the GC. The volatile pyrolysis products (pyrolyzate) are chromatographically separated by using a fused silica capillary column according to the boiling points and the affinity of analytes to the stationary phase (internal capillary column wall coating). The detection technique of the separated compounds is typically MS, but other GC detectors have also been used depending on the intentions of the analysis. The substances detected by the MS are subsequently identified by the interpretation of the obtained mass spectra, by using mass spectra libraries (e.g. NIST/EPA/NIH, Wiley, MPW, Norman Mass Bank, m/z Cloud), or by using reference substances. The identification of complex mixtures or blends as well as identification of samples with so-called "difficult matrices" is also possible in many cases. Due to these small sample amounts, the investigation of heterogeneous polymers with a coarse phase or a gradient composition structure (phase separation, poor mixing, etc.) is sophisticated and may lead to great variations in the measuring results. In this case, a multiple determination of different positions of the investigated part is essential to achieve a significant image of its composition [12].

2.1. Furnace Pyrolyzer

The furnace pyrolyzer is essentially an online inlet furnace which is continuously heated to the desired pyrolysis temperature. The sample is injected directly into the pyrolysis chamber using steel tubes, cups, or special types of syringes. An inert carrier gas (helium) flows through the pyrolysis chamber. The advantages of this system are that the pyrolysis conditions can be controlled very accurately and reproducible sample sizes are obtained more easily. In this system, condensation processes are reduced as well as memory effects [1]. In some systems, a cryogenic trap is used to focus the decomposition fragments. After the focusing phase, the trap is rapidly heated to desorb the products into the GC system. The major problem of this system is the poor contact between the sample and the hot source. In fact, the sample may reach a lower temperature than that of the furnace wall. Figure 2.4 shows the schematic diagram of a furnace pyrolyzer [13]. Figures 2.5 and 2.6 show the furnace pyrolyzer *Pyrojector II*™ (SGE Analytical Science, Melbourne, Australia) [14] and the micro-furnace pyrolyzer EGA/PY-3030D (Frontier Lab, Fukushima, Japan) [15], respectively.

For pyrolysis by using the *Pyrojector II*™ pyrolyzer, a plug of quartz wool should be packed and positioned in the center of the furnace (Fig. 2.5). This plug of quartz wool is required to prevent the sample from passing through the furnace into the cooler, lower part of the tube without undergoing satisfactory pyrolysis. The quartz wool also prevents particulate matter falling down into the transfer line, resulting in blockages. The desired pyrolysis temperature may be set from ambient temperature to 900°C in 1°C steps, and the carrier gas (helium) pressure is controlled by the use of an electronic control module.

Developed by Tsuge *et al.* [15–20], the multifunctional micro-furnace pyrolyzer (Fig. 2.6) for pyrolysis–GC and evolved gas analysis (EGA) of various synthetic and natural materials is composed of a double-shot

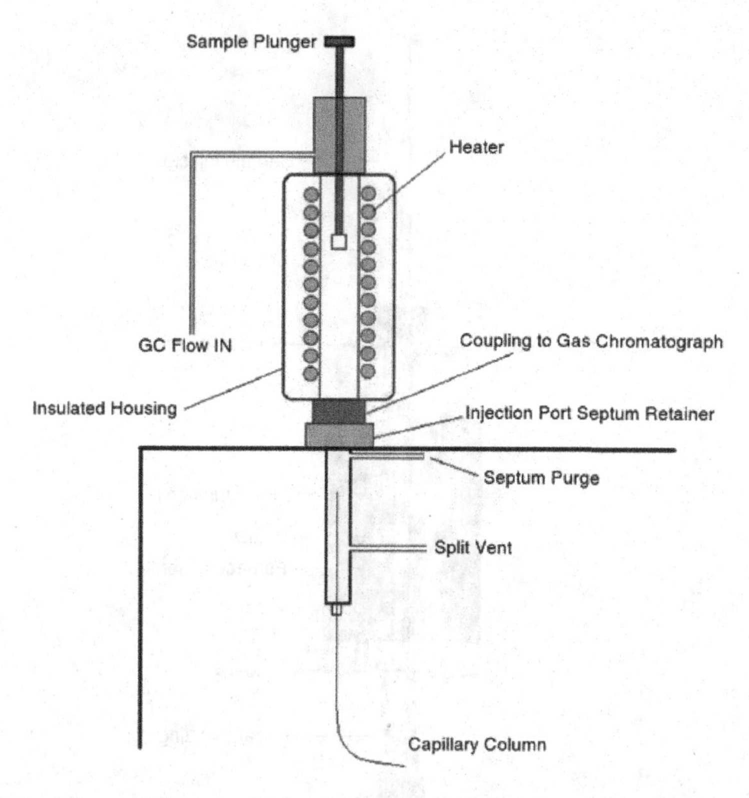

Fig. 2.4. Schematic diagram of a furnace pyrolyzer. Figure reprinted from Ref. [13] with permission from CRC Press.

micro-furnace with a sophisticated temperature control unit. The double-shot device enabled the thermal extraction of polymers at temperatures of 250–300°C for the determination of, e.g. plasticizers, flame retardants, or stabilizers in polymers following the pyrolysis of the sample. The newly developed multishot *Pyrolyzer*™ automatic sampler with a vertical furnace that employs free-fall sample introduction is suitable for the analysis of up to 48 samples [21]. Another furnace pyrolyzer based on the thermo-desorption unit (TDU) and using the automatic multipurpose sampler (MPS) was developed by the GERSTEL company (Fig. 2.7) for thermal extraction and freely selectable pyrolysis temperature up to 1000°C of liquid or solid samples [22].

Furnace pyrolyzers are inexpensive and relatively easy to use. Since they are operated isothermally, there are no controls for heating ramp

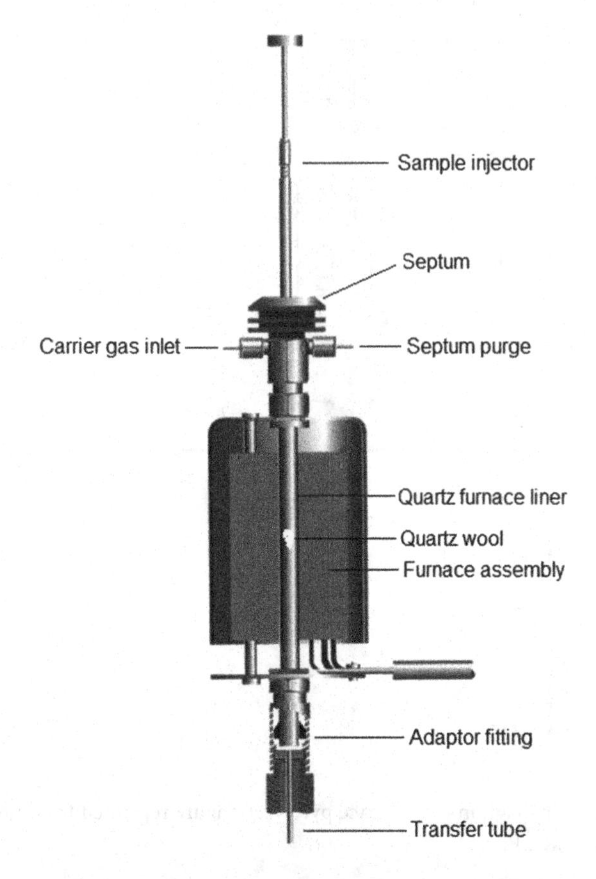

Fig. 2.5. Schematic view of the furnace pyrolyzer *Pyrojector II*™ (SGE Analytical Science, Melbourne, Australia) [14].

rate or pyrolysis time. Liquid and gaseous samples are pyrolyzed more easily than with filament-type pyrolyzers [13].

2.2. Curie-Point Pyrolyzer

The temperature at which ferromagnetic material is demagnetized by heating is physically defined as the Curie-point. After reaching the characteristic Curie-point, temperature stabilization is achieved by using ferromagnetic filaments of appropriate composition and a suitable induction frequency. The analytical Curie-point pyrolysis (CPPy) method was

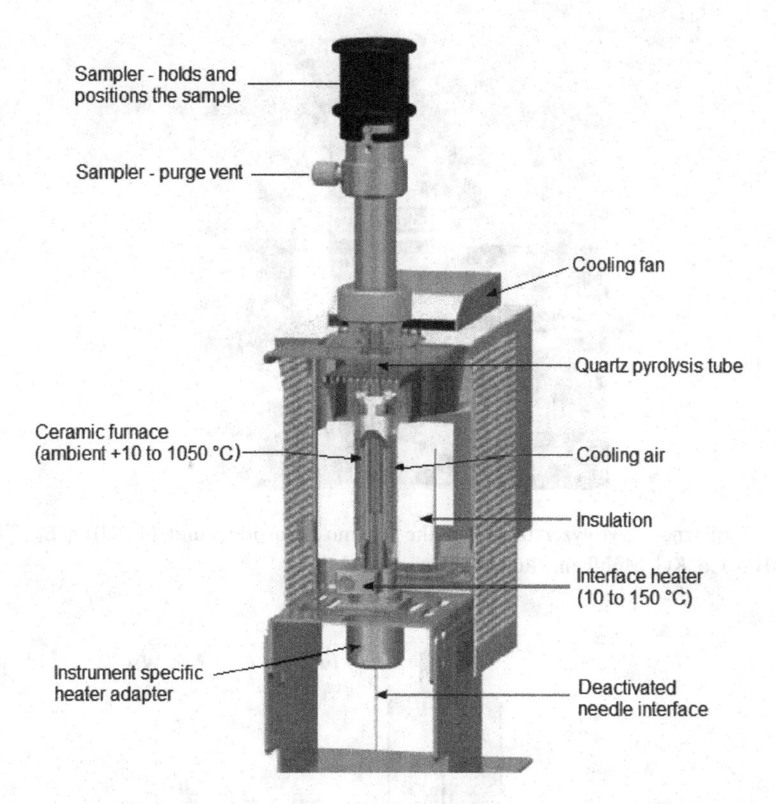

Fig. 2.6. Schematic view of the micro-furnace pyrolyzer EGA/PY-3030D (Frontier Lab, Fukushima, Japan) [15].

introduced by Simon and Giacobbo at the Eidgenössische Technische Hochschule (ETH), Zürich (Switzerland) in 1965 [23]. Figure 2.8 shows the schematic diagram of a Curie-point pyrolyzer. Figures 2.9 and 2.10 show the schematic views of the manual Curie-point pyrolyzer model 0316 (*Fischer* Labor- und Verfahrenstechnik GmbH, Meckenhein/Bonn, Germany) [1, 24–26] and the *Fischer* Curie-point device distributed today by GSG Mess- und Analysegeräte GmbH (Bruchsal, Germany), respectively.

The Curie-point technique used the inductive heating of filaments. In this technique, a ferromagnetic wire or foil is centered in a glass or quartz tube, which is connected to the injection port of a GC and through which the carrier gas (helium) flows (Fig. 2.9). A Curie-point induction coil surrounds the tube and the wire or foil is heated by induction. The wire or foil

Fig. 2.7. Furnace pyrolyzer based on the thermo-desorption unit (TDU) (GERSTEL GmbH & Co. KG, Mühlheim/Ruhr, Germany) [22].

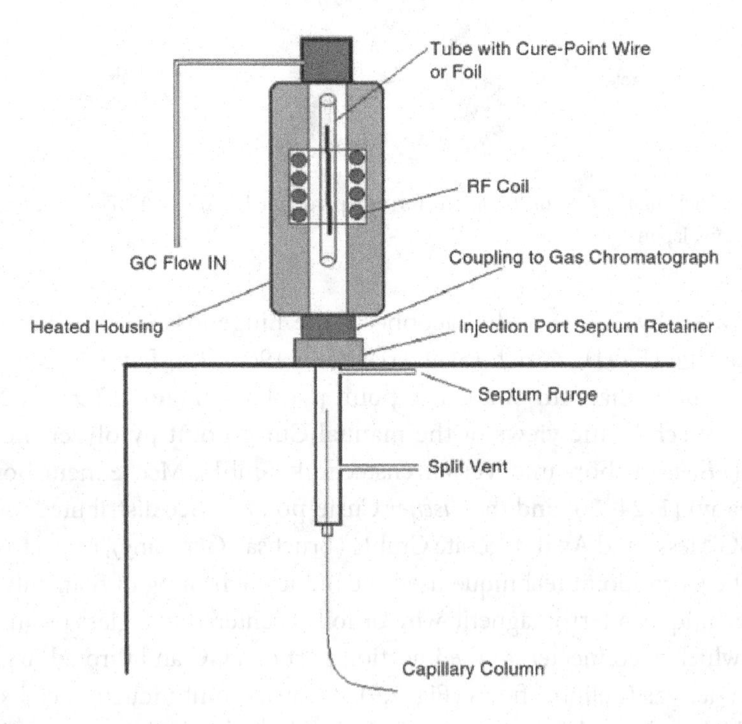

Fig. 2.8. Schematic diagram of a Curie-point pyrolyzer. Figure reprinted from Ref. [13] with permission from CRC Press.

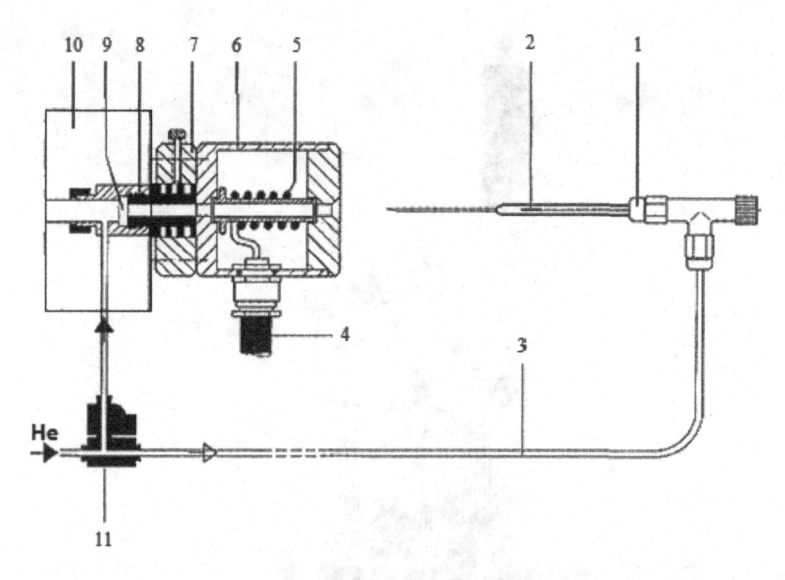

Fig. 2.9. Schematic view of the manual Curie-point pyrolyzer model 0316 (*Fischer* Labor- und Verfahrenstechnik GmbH, Meckenheim/Bonn, Germany) [1, 24–26]. 1, Pyrolysis injector made of glass with stainless steel injection needle; 2, ferromagnetic filament (sample support); 3, tetrafluoroethylene hose; 4, impulse cable; 5, induction coil; 6, aluminum housing; 7, adapter; 8, GC injector; 9, septum; 10, GC; 11, carrier gas switching valve. Carrier gas (helium) flow is indicated with arrows.

heats up until its Curie-point is reached. This occurs within 20–30 ms and is followed by subsequent stabilization of temperature. This is the temperature at which the wire becomes paramagnetic and its energy intake drops, thus holding the temperature of the wire or foil at this point. A range of pyrolysis temperatures is obtained by using alloys containing different amounts of the common ferromagnetic metals (e.g. iron, cobalt, nickel, and chromium). In Table 2.1 various Curie-point temperatures are shown, depending on the composition of the ferromagnetic metals in the alloy.

The basic equipment of the Curie-point pyrolyzers consists of a high-frequency oscillator with a ca. 2 kW impulse output, a built-in digital pyrolysis timer (0.1–10 s), a pyrolysis reactor with the induction coil, and the pyrolysis injector (manual or automatic) with the ferromagnetic sample carriers (filaments). The filaments are used in a wide range of shapes, such as straight wires, foils, spirals, and tubes. Solid samples of 5–200 μg can simply be placed within the bottom of a crimped tube,

Fig. 2.10. View of the *Fischer* automatic Curie-point pyrolyzer distributed by GSG Mess- und Analysegeräte GmbH (Bruchsal, Germany) [28].

whereas for solutions the filament can be dipped into the liquid and the solvent evaporated. The electromagnetic radio-frequency (RF) energy transfer from the oscillator to the ferromagnetic filament is done inductively. The pyrolysis temperature can only be varied by replacing the filament with one of a different alloy composition. The pyrolysis device with the automatic sampler (Fig. 2.10) consists of 12 or 24 replaceable quartz tubes with the ferromagnetic filaments arranged in a circular pattern. The quartz tubes with the samples placed on the sample supports (filaments) are put into the pyrolysis chamber and pyrolyzed in a programmed series. The outlet from the pyrolysis chamber is connected to the inlet port of the GC. Automatic triggering of the automatic sampler and GC is integrated within the system [27, 28].

A portable Curie-point system without complicated installation on a GC injector is provided by the JAI company (Fig. 2.11) [29]. The sample

Table 2.1. Dependence of the Curie-point on the composition of the ferromagnetic metal in the alloy.

Composition of the ferromagnetic metals in the alloy (%, m/m)			
Iron	Cobalt	Nickel	Curie-point (°C)
0	100	0	1075
0	60	40	900
0	33	67	660
0	0	100	358
33.3	33.3	33.3	700
42	42	16	600
50	50	0	950
50	0	50	470
60	0	40	320
61.7	28.3	0	400
100	0	0	770

is enclosed in a ferromagnetic filament, pyrofoil, or deposited on ferromagnetic wire, pyrowire, which is placed in a flow path of a GC carrier gas (Fig. 2.11). The foil or the wire is rapidly heated to the Curie-point of the ferromagnetic filament with the induction of RF. The Curie-point is reached instantaneously, in less than 0.2 s. Twenty-one pyrofoils and ten pyrowires, each with distinct Curie-point temperature between 160°C and 1040°C, are available for analyzing wide range of compounds. Sample tube and needle are washable to keep the flow line and the GC injection port clean [29]. The manufacturer also offers a hybrid pyrolyzer. The CPPy and the furnace pyrolysis are both incorporated in one system (Fig. 2.12) [29].

The most important features of the Curie-point pyrolyzers are the exactly reproducible temperatures of pyrolysis, "shock-like" heating-up within milliseconds, and a wide temperature range due to the presence of ferromagnetic filaments or tubes from 250°C to 1200°C [25–27]. The disadvantage of the Curie pyrolyzers is that not any pyrolysis temperature is applicable. The pyrolysis temperature depends on the composition of the ferromagnetic metals in alloys used for filaments.

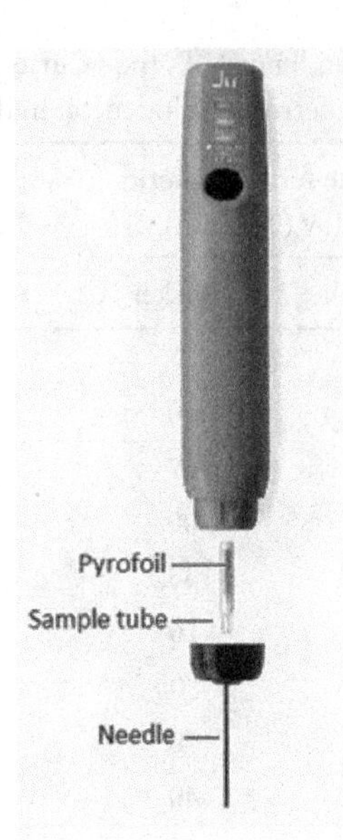

Fig. 2.11. View of the portable Curie-point pyrolyzer (JAI Japan Analytical Industry Co., Ltd., Tokyo, Japan) [29].

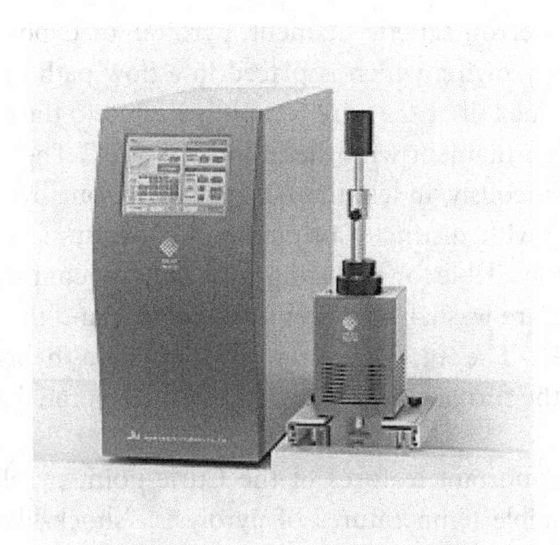

Fig. 2.12. View of the hybrid pyrolyzer (JAI Japan Analytical Industry Co., Ltd., Tokyo, Japan) [29].

2.3. Resistively Heated Filament Pyrolyzer

Like the Curie-point instruments, resistively heated filament pyrolyzers operate by taking a small sample from ambient to pyrolysis temperature in a very short time. Current supply is connected directly to the filament, however, and not included [13]. This means that the filament does not need to be ferromagnetic, but it must be physically connected to the temperature controller of the instrument. Filaments are generally made of materials of high electrical resistance and wide operating range and include iron, platinum, and nichrome [13].

The principle of this type of pyrolyzer is that an electric current passing through a resistive conductor generates heat in accordance with Joule's law:

$$Q = I^2Rt = \frac{V^2t}{R}, \tag{2.1}$$

where Q is the amount of heat (in J), I is the current intensity (in A), R is the electrical resistance of the conductor (in Ohms), t is the time in seconds, and V is the voltage (in V) [30]. A simple flash pyrolysis unit that operates at a fixed voltage could easily be constructed. However, such a unit operating within common values for the current intensity and voltage will have a temperature-rise time (TRT) that is too long to be appropriate for flash pyrolysis. This parameter measures the time necessary for the heating element of the pyrolyzer to reach the equilibrium temperature (T_{eq}). The equilibrium temperature is the targeted isothermal condition for flash pyrolysis. Systems with boosted current or boosted voltage were used to achieve a more rapid heating [30]. These systems apply a constant current to the filament with an initial boost pulse to assure a rapid temperature increase at the beginning of the heating period. For a filament pyrolyzer, TRT values lower than 7 ms from ambient to 1000°C were reported [30]. Figure 2.13 shows a schematic view of a resistively heated pyrolyzer installed on a GC injection port [13].

The CDS *Pyroprobe®* (CDS Analytical, Oxford, PA, USA) model series 5000 and 6000 pyrolyzers used the platinum filaments (Fig. 2.14) for rapidly heated pulse pyrolysis work or for slowly heated with controlled rates for programmed analyses [31]. The pyrolysis temperatures up to 1400°C

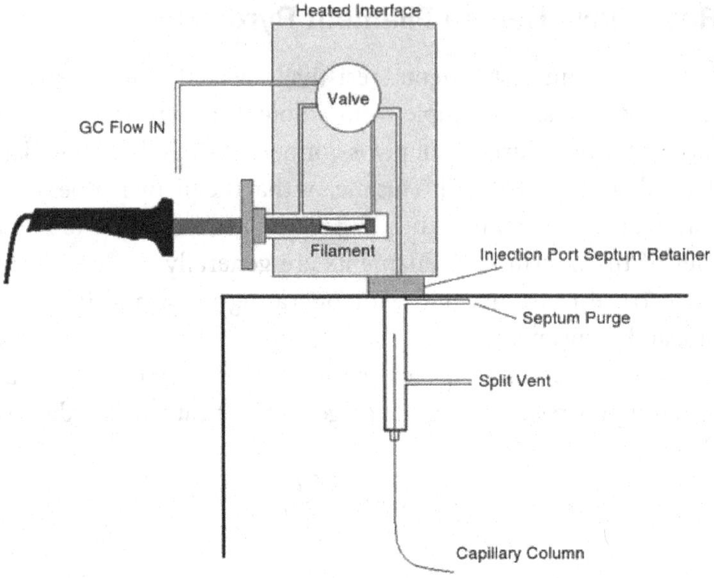

Fig. 2.13. Resistively heated filament pyrolyzer. Figure reprinted from Ref. [13] with permission from CRC Press.

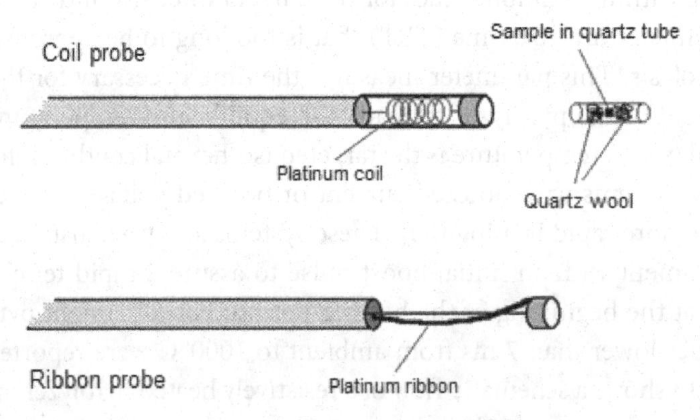

Fig. 2.14. Schematic view of the *Pyroprobe*® filament rods (CDS Analytical, Oxford, PA, USA).

can be reached in 1°C increments for a wide heating range. Analytical runs may be programmed for up to eight steps per sample, with automatic control of the online valve, interface temperature, GC ready sensing, and GC start for each step [31]. When configured with the trapping option, the *Pyroprobe*® pyrolyzers may be used to collect analytes from slow rate pyrolysis, thermal desorption, or reactant gas pyrolysis. The interfacing design permits a direct pyrolysis path to the GC inlet or rapid sample heating and transfer to the trap without interruption of the pneumatics of the GC [31].

In the filament pulse-mode pyrolyzer *Pyrola*® *2000* (Pyrol AB, Lund, Sweden), based on the research at the Department of Analytical Chemistry of the University of Lund (Sweden) [32–36], the platinum filament is heated resistively by two current pulses. The sample that is to be pyrolyzed is placed on a thin metal foil ($15 \times 2.6 \times 0.012$ mm^3). The foil is usually made of platinum, but palladium, molybdenum, iron, iron/nickel, and nickel foils have also been used [36]. The TRT can thus be very short — a few milliseconds up to 1400°C [36]. As a complement to isothermal pyrolysis, *Pyrola*® *2000* includes three pyrolysis techniques: sequential pyrolysis, fractionated pyrolysis, and pyrotomy [36]. In sequential pyrolysis, the sample is heated repeatedly to the same temperature, which is set sufficiently low to maintain a sufficient amount of sample for the subsequent pyrolysis step. The results can be used for determining the thermal degradation rate of the sample and can also be used for qualitative analyses, e.g. to distinguish between a homopolymer and copolymer. In fractionated pyrolysis, the sample is heated repeatedly, but to different temperatures. It is therefore possible to separate substances with different degradation rates. Thus, fractionated pyrolysis is especially suited for the analysis of complex unknown samples [36]. In pyrotomy, the sample is exposed to several extremely short thermal pulses (on the order of milliseconds). Only the part of the sample that is in direct contact with the platinum filament will be heated in each pulse, giving a separate pyrolysis of each layer of the sample. Therefore, if the sample consists of a laminate, the pyrograms will give information about each layer separately, instead of having all of them mixed in a single pyrogram [36]. Figure 2.15 shows the manual filament pulse-mode pyrolyzer *Pyrola*® *2000* (Pyrol AB, Lund, Sweden).

Fig. 2.15. View of the manual filament pulse-mode pyrolyzer with control unit *Pyrola® 2000* (Pyrol AB, Lund, Sweden) [36].

The main advantage of a resistively heated pyrolyzer is that the filament may be heated to any temperature over its usable range, at a variety of rates. This permits the examination of a sample material over a range of temperatures without the need to change filaments for each temperature [13]. The main disadvantage of a resistively heated pyrolyzer results from the fact that the filament must be physically connected to the controller. The temperature control of a resistively heated filament is based on the resistance of the entire filament loop, including the filament and its connecting wires. Anything, that damages or alters the resistance of any part of the loop will have an effect on the actual temperature produced by the controller [13].

An additional disadvantage may be produced by the fact that the filament must be housed in a heated zone. The introduction of some samples into a heated chamber before pyrolyzing them may produce volatilization or denaturation, altering the nature of the sample before it is actually degraded [13].

2.4. Laser Pyrolyzer

In laser pyrolysis, the interaction of laser energy with matter generates a high-temperature plume. A traditional drawback of laser pyrolysis has been that some samples, particularly those that are transparent, do not couple efficiently with the laser radiation [1, 37, 38]. Previous researchers have incorporated absorption centers in the sample by adding powdered carbon or nickel or by depositing the sample on an absorbing surface, such as blue cobalt glass, to increase interaction between the sample and the laser radiation [37]. An instrument for rapid characterization of synthetic polymers using a UV laser coupled to fast chromatography and time-of-flight mass spectrometry (TOF-MS) was presented by Meruva, Metz, Goode, and Morgan from the University of South Carolina, Columbia, SC, USA [37, 38]. The instrument consists of a UV laser source, a sample cell for pyrolysis, a valve interface system, and a fast GC/TOF-MS. A *Q*-switched Nd:YAG laser, frequency quadrupled to 266 nm, was used as the fragmentation source for the degradation of polymers. The flat beam profile of the laser enables reproducible analysis of thin-layer sample surfaces. The sample cell is constructed of stainless steel with a replaceable laser-grade quartz window and has a total volume of 5.27×10^3 mm^3 in the closed position. Pyrolysis products are swept by helium through a heated transfer line directly to the valve interface, and subsequently to the fast GC [37, 38].

Instruments for laser micropyrolysis–GC/MS were also presented by Greenwood *et al.* [39] and da Silva *et al.* [40]. Table 2.2 summarizes the main characteristics of several pyrolyzers.

Table 2.2. Comparison of the main characteristics of several pyrolyzers [41].

Property	Curie-point	Heated filament	Micro-furnace	Laser
Temperature limit (°C)	1200	1400	1050	High
Temperature control	Discrete	Continuous	Continuous	Uncontrolled
Use of temperature gradients	Not possible	Possible	Common	Possible
Minimum TRT	20–30 ms	10 ms	0.2 s to 1 min	10 μs
Sample size (μg)	10–1000	10–1000	50–5000	20–500
Reproducibility	Very good	Very good	Good	Poor
Catalytic reactions	Some	Low	Low	Very low
Autosampler availability	Yes	Yes	Yes	No
Use with analytical instruments	Online/offline	Online/offline	Online/offline	Online/offline

3
Degradation Mechanisms of Polymers

Starting at the University of Cologne (Germany) in the 1960s, Hummel and co-workers studied the process of thermal decomposition of polymers in detail using Py–GC/MS and Py–GC/FTIR [9–10]. In general, decomposition proceeds through radical formation, which, due to the high reactivity of radicals, initiates numerous consecutive and parallel reactions. The authors summarized the main pathways of polymer decomposition in four categories:

1. retropolymerization from the end of the polymer chain, predominantly forming monomers (e.g. poly(methyl methacrylate) (PMMA), poly(α-methylstyrene));
2. statistical chain scission followed by:
   — retropolymerization from radical bearing chain ends (e.g. poly-isobutylene (PIB), polystyrene (PS));
   — radical transfer and disproportionation (e.g. polyethylene (PE), isotactic polypropylene (PP)); and
   — stabilization of fragments by cyclization (e.g. polydimethylsiloxane (PDMS)).
3. splitting side chain leaving groups (e.g. poly(vinyl chloride) (PVC), polyacrylonitrile, polyacrylates); and
4. intramolecular condensation reactions with loss of smaller molecules (e.g. phenol-epoxide resins).

However, this classification is restricted to homogeneous polymers. The situation is more complex in copolymers, depending on the applied monomers and their linking [9, 10].

In spite of the high complexity of the pyrolysis reactions there is still the possibility to obtain information on the chemical composition of the parent compound. Some cleavages of specific bonds are quite well predictable due to their lower stability. These occur due to the relative strengths of the chemical bonds between the atoms within the macromolecular material. In general, stable molecules containing strong bonds need higher pyrolysis temperatures to initiate cleavages than weaker bonds, which crack at lower energies [41, 42].

There are different reaction types depending on the complexity of the whole molecule that can occur during pyrolysis. Some of those are the elimination reaction (α-, β-, 1,3- and 1,*n*-elimination), the fragmentation reactions (fragmentations, retro-ene reactions, retro-Diels-Alder reactions, condensations, etc.), the rearrangements (1,2-migrations, electrocyclic rearrangements, etc.), and others [42].

3.1. Elimination Reactions

In organic chemistry there are two different types of elimination reactions. These are referred to as E_1 and E_2 where "E" stands for elimination, "1" for unimolecular, and "2" for bimolecular. In the E_1 reaction the base or nucleophil (B^-) does not have to be involved, whereas in E_2 the base or nucleophil must be part of the mechanism by pulling the proton. The loss of the leaving group and the removal of the proton happen successively during E_1 reactions but concerted in E_2 mechanisms [42]. Figure 3.1 shows two schematic reaction mechanisms for both E_1 and E_2 types of elimination [41, 42].

In the E_1 mechanism, the reagent loses an X^- group to form a carbocation. In the next step a proton is lost to a (Lewis) base. In the E_2 mechanism a proton is pulled by the base and the X^- group departs simultaneously. During pyrolysis elimination is a common reaction, which probably dominates most pyrolysis processes. However, typical E_1 and E_2 reactions are not common in pyrolysis in gas phase [30, 41]. As pyrolysis does not need any further reagents but thermal energy, Moldoveanu named this type of

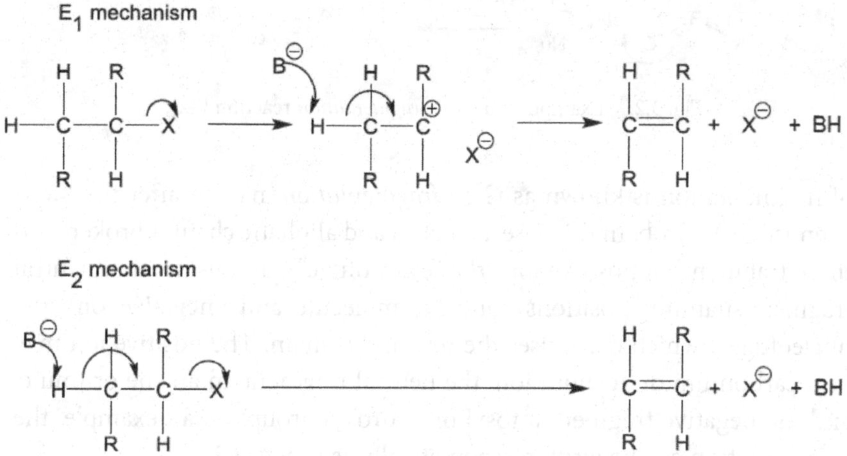

E_1 mechanism

E_2 mechanism

Fig. 3.1. Schematic reaction mechanisms for both E_1 and E_2 types of eliminations [41, 42].

elimination reaction as "E_i" mechanism, where "i" stands for internal [30, 43]. The E_i mechanism, which is very common in pyrolysis, involves a cyclic transition state, which may be four-, five-, or six-membered states [41].

Especially molecules with bonds of similar strengths cleave randomly, i.e. alkane chains. Elimination reactions can be classified by the position of the atoms involved as α-elimination, β-elimination, 1,3-eliminations, etc. α-Eliminations involve two leaving groups connected to the same carbon (an α-carbon is the carbon that attaches a functional group). They are encountered in some pyrolytic reactions where the more common β-elimination is not possible. β-Elimination is the most common reaction in pyrolysis by which two groups are lost from adjacent atoms. This reaction commonly involves an E_i mechanism (ring transition state and loss of two vicinal groups in *syn* positions). Besides β-elimination, 1,3- or 1,*n*-eliminations also may take place during pyrolysis with the formation of cycles. One reaction of this type takes place during the pyrolysis of nylon 66 (see Section 4.15) [30].

3.2. Fragmentation Reactions

In fragmentation reactions a molecule (e.g. A–B) is cleaved into two parts: A and B. The bonds between the fragments are rearranged. This cleavage leads to the formation of either ions or radicals [41, 42]. One type

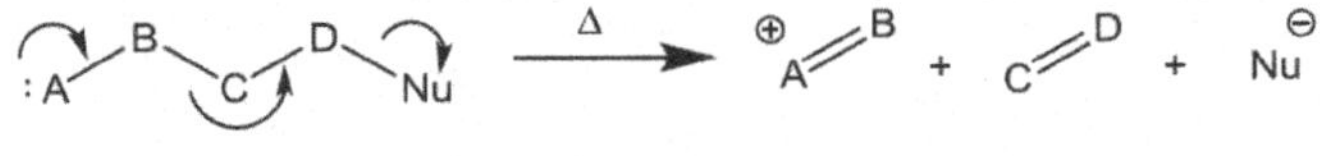

Fig. 3.2. Example of a *Grob fragmentation* reaction [42].

of fragmentation is known as *Grob fragmentation*, named after the Swiss chemist C. A. Grob. In this case, a neutral and aliphatic chain is broken into three fragments: a positive ion (the "electrofuge"), an unsaturated neutral fragment spanning positions 3 and 4 in molecule, and a negative ion ("the nucleofuge") which comprises the rest of the chain. The positive ion may be a carbonium or acylium ion, the neutral fragment an alkene or imine, and the negative fragment a tosyl or hydroxyl group. As an example, the fragmentation can be written schematically as in Fig. 3.2.

The formed radicals further react with other ions to build stable molecules. This mechanism can be concerted or occur as a two-step process in which the order of the leaving fragment (electrofuge or nucleofuge first) is the decisive factor for the formed transition-state products.

3.3. Rearrangement Reactions

Another type of reaction in pyrolysis is the *rearrangement*. Rearrangements are common during pyrolysis following eliminations (mainly when radicalic mechanisms are involved) and fragmentations. Rearrangement mechanisms take place when a group migrates from one atom to another within the same molecule. There are several types of rearrangement reactions known (e.g. electrocyclic, sigmatropic, 1,2-migration), whereby the 1,2-migration is common for aryl, vinyl, acetoxy, and halogen groups [41, 42].

3.4. Chain Scission

Polymer degradation reactions are frequently based on the site in the macromolecule structure where the reaction occurs. This leads to the following classification of scission reactions [30]:

- polymeric chain scission;
- side group reactions; and
- combined reactions.

These reactions follow one of the mechanisms described previously. The polymeric chain scission is an elimination reaction. The reaction occurs by breaking the bonds from the polymeric chain. This reaction may take place as a successive removal of the monomer units (end cleavage) or may occur as a random cleavage of the polymeric chain. The random cleavage reaction mainly takes place when the bonding energy is similar along the chain [30]. Many chain scission reactions have a free radical mechanism. Detailed description of degradation mechanisms of different classes of polymers can be found in Moldoveanu's monography [30].

4

Pyrolysis–Gas Chromatography/ Mass Spectrometry of Different Classes of Synthetic Polymers and Copolymers

Since the beginning of the 20th century a rapid development has taken place in the field of macromolecular chemistry, the end of which has not yet been foreseen. Polymeric commodities and specialty products are widely used in modern life. Today, there is a large number of plastics on the market which are compounded according to the application. All of these products are complex mixtures where compositions of both, the actual polymers and the additives, have been carefully fine-tuned to obtain the desired properties [44]. Additives include antioxidants, chemical stabilizers, bio-stabilizers, light-stabilizers, crosslinkers, lubricants, processing aids, impact modifiers, fillers, fire retardants, antistatic agents, optical brighteners, chemical propellants, and plasticizers. Additives can improve or modify the mechanical properties (fillers and reinforcements), modify the color and appearance (pigments and dyestuffs), give resistance to heat degradation (antioxidants and stabilizers), provide resistance to light degradation (UV stabilizers), improve the flame resistance (flame retardants), improve the performance (antistatic or conductive additives, plasticizers, blowing agents, lubricants, mold release agents, surfactants, and preservatives), and improve the processing characteristics (recycling additives) of polymers or copolymers [45–47]. Due to the large number of substances involved, the chemical identification of a plastic becomes a complex task. Detailed information of the chemical structure, molecular weight (MW) and the levels of the additives in the polymer is crucial for understanding and improving the properties of a product. It is for this reason that the development of analytical tools for such analyses has received a great deal of attention. The most important compositional parameters of polymeric materials are the composition of the polymer phase on the one hand, and of the additive package on the other hand [44]. Even for the simplest polymeric systems a full characterization is not trivial. Several distributions will generally be present in the polymer, i.e. distributions of chain length, degree of branching, end-groups, etc. For mixtures of homopolymers or for copolymers this is even more complex. Particularly difficult, but at the same time also highly relevant, is the determination of copolymer composition as a function of MW [44].

Plastics can be divided into two major categories: biopolymers and synthetic polymers.

Biopolymers are polymers synthesized by living organisms. Biopolymers can be polynucleotides (such as the nucleic acids DNA and RNA), polypeptides (proteins), or polysaccharides (polymeric carbohydrates). These consist of long chains made of repeating, covalently bonded units, such as nucleotides, amino acids, or monosaccharides. There are four types of biopolymer materials: sugar-based biopolymers, (2) starch-based biopolymers, (3) cellulose-based biopolymers, and (4) synthetic materials-based biopolymers.

Synthetic polymers can be divided into (1) thermoset or thermosetting plastics, (2) thermoplastics, and (3) elastomers. Thermosets are hard and durable. Once cooled and hardened, these plastics retain their shapes and cannot return to their original form. Examples include polyurethanes (PU), polyimides (PI), epoxy resins (EP), aminoplasts (UF, MF), and phenolic resins (PF). Thermosets can be used for auto parts, aircraft parts, and tires. Thermoplastics are less rigid than thermosets and can soften upon heating and return to their original form. They are easily molded and extruded into films, fibers, and packaging. Examples include PE, PP, PVC, polyethylene terephthalate (PET), PS, polyacrylates (PMMA, PAN), polytetrafluoroethylene (Teflon), or polyamides (PA).

Elastomers are a group of lightly crosslinked polymers that exhibit elastic or viscoelastic deformation. Thermal analysis (TA) and analytical pyrolysis (Py) play an important role in the analysis of elastomers. They are widely used to characterize raw materials, intermediate products, and vulcanization products. The information obtained is valuable for quality control, process optimization, research and development of advanced materials, and failure analysis. Examples of elastomers include ethylene–propylene–diene rubber (EPDM), styrene–butadiene rubber (SBR), nitrile-butadiene rubber (NBR), and ethylene–vinyl acetate copolymer (EVA).

4.1. Instrumentation and Method

Approximately 100–200 μg of solid polymer/copolymer sample were cut out with a scalpel and inserted into the bore of the pyrolysis solids-injector, without any further preparation, and then placed on the quartz wool of the quartz tube of the furnace pyrolyzer *Pyrojector II*™ (SGE Analytical Science, Melbourne, Australia) (Figs. 2.2, 2.3, and 2.5) with the plunger. The pyrolyzer was operated at a constant temperature of 550°C, 600°C, or 700°C. The pressure of helium carrier gas at the inlet to the furnace was 95 kPa. Pyrolysis–GC/MS measurements were made by using two apparatuses. In the first apparatus (1) the pyrolyzer was connected to a *7890A* GC with a series *5975C* quadrupole MS (Agilent Technologies Inc., Santa Clara, CA, USA) operated in electron impact ionization (EI) mode (Fig. 2.2). Two fused silica capillary columns: (1) 60-m long, 0.25-mm ID or (2) 59-m long, 0.25-mm ID with DB 5-ms stationary phase, film thickness 0.25-μm were used (J&W Scientific, Folsom, CA, USA). For some experiments a new capillary column (1 UI) 60-m long, 0.25-mm ID with DB 5-ms UI stationary phase, film thickness 0.25-μm was used. Helium, grade 5.0 (Westfalen AG, Münster, Germany) was used as a carrier gas. The GC conditions were as follows:

(1) programmed temperature of the capillary column from 75°C (1-min hold) at 7°C/min to 280°C (hold to the end of analysis) and programmed pressure of helium from 122.2 kPa (1-min hold) at 7 kPa/min to 212.9 kPa (hold to the end of analysis), and

(2) programmed temperature of the capillary column from 75°C (1-min hold) at 7°C/min to 280°C (hold to the end of analysis) and constant helium flow of 1 cm^3/min during the whole analysis.

The temperature of the split/splitless injector was 250°C and the split ratio was 50:1. The transfer line temperature was 280°C. The EI ion source temperature was kept at 230°C. The ionization occurred with a kinetic energy of the impacting electrons of 70 eV. The quadrupole temperature was 150°C. Mass spectra and reconstructed chromatograms [total ion current (TIC)] were obtained by automatic scanning in the mass range *m/z* 35–750 u. GC/MS data were processed with the *ChemStation* software (Agilent Technologies) and the *NIST 05* mass spectra library.

In the second apparatus (2), the pyrolyzer *Pyrojector II*™ was connected to a *Trace 2000* GC (ThermoQuest/CE Instruments, Milan, Italy) with a quadrupole MS *Voyager* (ThermoQuest/Finnigan, MassLab Group, Manchester, UK) operated in electron impact ionization (EI) mode (Fig. 2.3). The fused silica GC capillary columns (3) Elite 5-ms 60-m long, 0.25-mm ID, 0.25-μm film thickness (PerkinElmer Instruments, Shelton, CT, USA) and (4) BPX-50, 60-m long, 0.25-mm ID, 0.25-μm film thickness (SGE Analytical Science, Melbourne, Australia) were used. The GC conditions were as follows:

(3) programmed temperature of the capillary column from 60°C (1-min hold) at 2.5°C/min to 100°C and then 10°C/min to 280°C (hold to the end of analysis), and
(4) programmed temperature of the capillary column from 60°C (7-min hold) at 5°C/min to 100°C and then 10°C/min to 280°C (hold to the end of analysis).

The temperature of the split/splitless injector was 250°C and the split flow was 50 cm^3/min. Helium, grade 5.0 (Westfalen AG, Münster, Germany) was used as a carrier gas at programmed pressure of 70 kPa (1-min) and then 1 kPa/min to 110 kPa and then constant at 110 kPa to the end of the analysis. The transfer line temperature was 280°C. The MS EI ion source temperature was kept at 250°C. The ionization occurred with a kinetic energy of the impacting electrons of 70 eV. The current emission of the rhenium filament was 150 μA. The MS detector voltage was 350 V. Mass spectra and reconstructed chromatograms (TIC) were obtained by automatic scanning in the mass range *m/z* 35–450 u. Pyrolysis–GC/MS data were processed with *Xcalibur* software (ThermoQuest) and the *NIST 05* mass spectra library.

In the third apparatus (3), the pyrolyzer *Pyrojector II*™ was connected to a *Clarus 500* GC (PerkinElmer Instruments, Shelton, CT, USA) with a split/splitless injector and a flame ionization detector (FID). The fused silica capillary column (5) 60-m long, 0.25-mm ID with DB 5-ms stationary phase, film thickness 0.25-μm (J&W Scientific, Folsom, CA, USA) was used. The GC conditions were as follows:

(5) programmed temperature of the capillary column from 75°C (1-min hold) at 7°C/min to 280°C (hold 15 min).

The temperature of the split/splitless injector was 250°C and the split flow was 50 cm^3/min. Helium, grade 5.0 (Westfalen AG, Münster, Germany) was used as a carrier gas at constant pressure of 120 kPa. The temperature of the FID was 280°C.

4.2. Analysis of Polyolefins

Polyolefins (polyalkenes) are thermoplastic hydrocarbon polymers formed from a chain of aliphatic carbon atoms. This class includes polymers such as PE, PP, poly(1-butene) (polybutylene, PB), PIB, poly(4-methyl-1-pentene) (PMP), etc. Polyolefins are the most important plastic group. If the hydrogen atom is replaced by a halogen atom in these polymers, the polyhaloolefins are formed. The main representatives are poly(teterafluoroethylene) (PTFE) when using fluorine, and PVC when using chlorine. High-density polyolefins are crystalline, whereas the low-density polyolefins are amorphous. Analytical pyrolysis has been widely used in the study of polyolefins in general.

4.2.1. Characterization of polyethylene (PE)

PE has the simplest structural unit, *viz.* $-CH_2-CH_2-$. The intermolecular attraction in PE is low due to the absence of polar or polarizable groups, but the high symmetry of the molecules and narrow chain width facilitate crystallization so that in the best crystallized sample the melting temperature is *ca.*130°C [48]. However, PE does not have the ideal linear structure, $-(CH_2-CH_2)_n-$. Various kinds of PE exist, which differ in microstructures. They are differentiated according to density as low-density polyethylene (LDPE, ρ = 0.910–0.935 g/cm^3), high-density polyethylene (HDPE, ρ = 0.941–0.965 g/cm^3), and linear low-density polyethylene (LLDPE, ρ = 0.87–0.94 g/cm^3). The density difference is due to the difference in the degree of crystalline order, which is dependent on the microstructure resulting from the method of synthesis. LDPE is highly branched with most of the branches being short alkyl groups but some being long chains. LLDPE has only short alkyl groups as branches, commonly made by copolymerization of ethylene with short-chain α-olefins, like 1-butene, 1-hexene, and 1-octene. HDPE is only lightly branched with short branches [48]. The microstructure of each of these three classes of PE is shown schematically in Fig. 4.1.

PE is highly versatile and probably the most used plastic material in daily life. The areas of applicability are still in continuous expansion to embrace a wide range of use and therefore the analysis of PE has been of

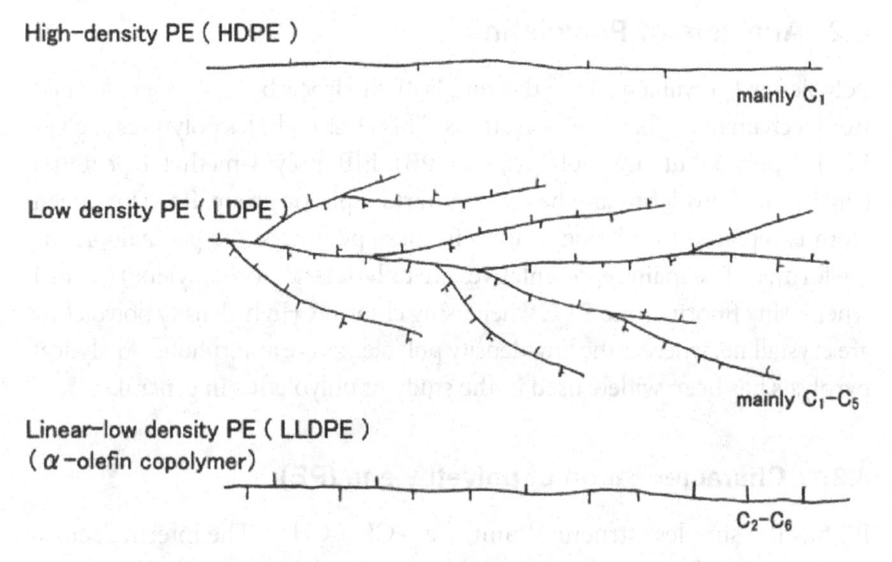

Fig. 4.1. Probable branching structures for various PEs. Figure reprinted from Ref. [49] with permission from CRC Press.

importance since many decades [50]. Pyrolysis of PE occurs through a random scission mechanism [13, 30, 50–52]. The carbon backbone is broken to produce a wide range of smaller hydrocarbons with terminal free radicals, which may be stabilized either by abstraction of hydrogen or beta scission [13, 30, 50–52]. If the free radical abstracts a hydrogen atom from a neighbouring molecule, it becomes a saturated end-group, creating a free radical in the neighbouring molecule. Beta scission results in degradation of the polymer backbone, producing an unsaturated end and a new terminal free radical. This process continues to produce hydrocarbon molecules, which are saturated (*n*-alkanes), have a double bond (α-alkenes), or have a double bond at each end (α,ω-alkadienes) [13, 30, 50–52]. The theoretical model for the PE thermal cracking mechanism and the formation of *n*-alkanes, α-alkenes, and α,ω-alkadienes by breaking the ß-carbon bonds based on the publication of González-Pérez *et al.* [50] is shown in Fig. 4.2. Figure 4.3 shows a typical Py–GC/MS TIC chromatogram obtained by pyrolysis of commercially available PE packaging at 700°C by use of the furnace pyrolyzer. The pyrogram consists of serial triplets, corresponding to C_3-C_{37} α,ω-alkadienes, α-alkenes, and *n*-alkanes, respectively, in the order of increasing $n + 1$ carbon number in the molecule. Identification of compounds was carried out by comparison of retention times and mass spectra of standards, study of the mass spectra, and comparison with data in the *NIST 05* mass spectra library. An example of the identified C_9–C_{12}-hydrocarbons triplets is presented in Fig. 4.4. By using the nonpolar GC stationary phase (see Section 4.1), the elution order of compounds in hydrocarbons triplets is as follow: α,ω-alkadienes $<$ α-alkenes $<$ *n*-alkanes.

4.2.2. Characterization of polypropylene (PP)

PP is a crystalline thermoplastic and one of the major members of the polyolefins family. It is a synthetic, high-molecular mass, linear, addition polymer of propylene. PP can be classified depending on the orientation of each methyl group ($-CH_3$) relative to the methyl groups on neighbouring monomers. This orientation has a strong effect on the finished polymer's ability to form crystals, because each methyl group takes up space and constrains backbone bending [53].

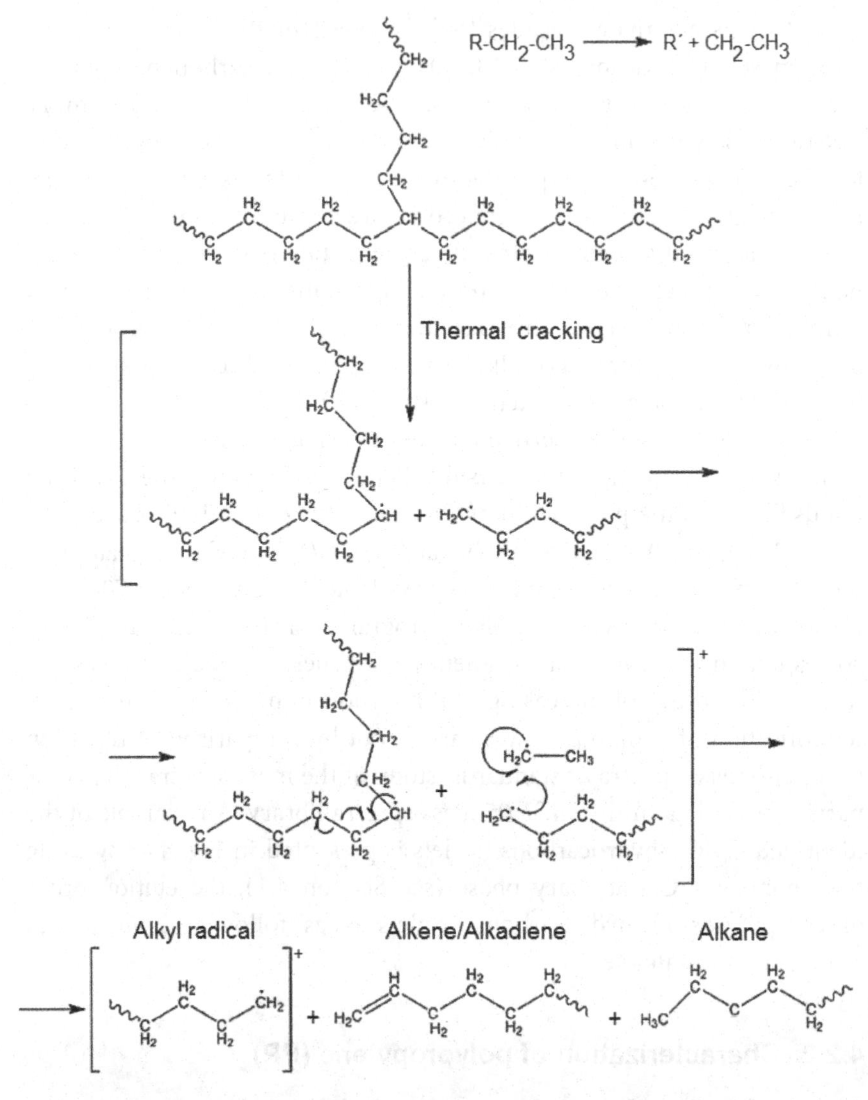

Fig. 4.2. Theoretical model mechanism for the PE thermal cracking and the formation of *n*-alkanes, α-alkenes and α,ω-alkadienes by breaking the ß-carbon bonds (based on Ref. [50] with permission from Elsevier).

Isotactic PP is formed by branched monomers that have the characteristic of having all the branch groups on the same side of the polymeric chain. The monomers are all oriented in the same way [53].

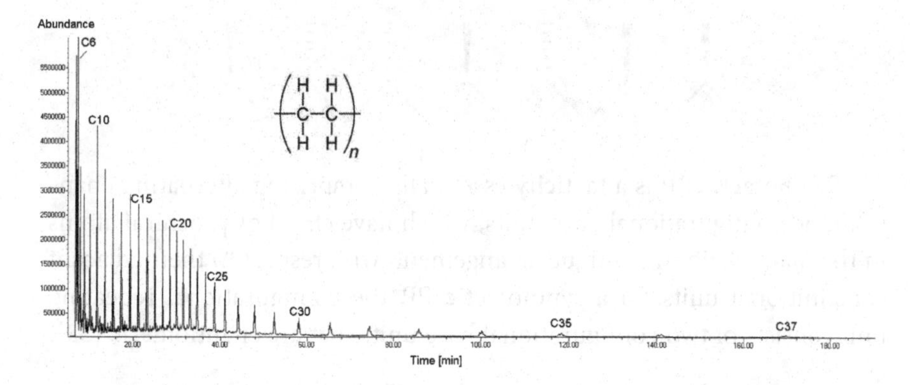

Fig. 4.3. Pyrolysis–GC/MS chromatogram of PE at 700°C. Apparatus 1, GC column 1, GC conditions 1. For peak identification, see text.

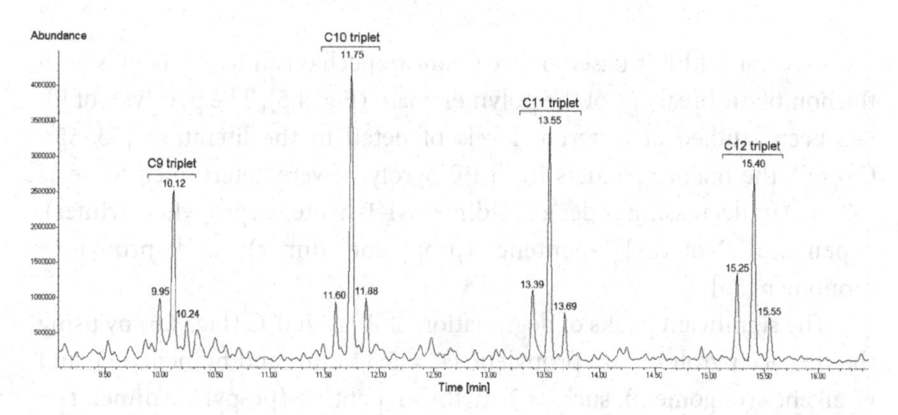

Fig. 4.4. Pyrolysis–GC/MS chromatogram of PE at 700°C, range 9.1–16.4 min. Apparatus 1, GC column 1, GC conditions 1. Peak identification — C9 triplet: t_R = 9.95 min — 1,8-nonadiene, t_R = 10.12 min — 1-nonene, t_R = 10.24 min — *n*-nonane; C10 triplet: t_R = 11.60 min — 1,9-decadiene, t_R = 11.75 min — 1-decene, t_R = 11.88 min — *n*-decane; C11 triplet: t_R = 13.39 min — 1,10-undecadiene, t_R = 13.55 min — 1-undecene, t_R = 13.69 min — *n*-undecane; C12 triplet: t_R = 15.25 min — 1,11-dodecadiene, t_R = 15.40 min 1-dodecene, t_R = 15.55 min — *n*-dodecane.

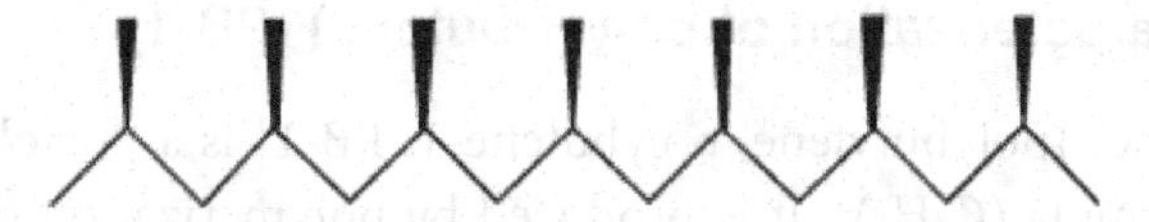

In atactic PP the $-CH_3$ substituent belonging to a repeating unit is placed randomly at either side of the backbone [53].

Syndiotactic PP is a tacticity essentially comprising alternating enantiomeric configurational base units, which have chiral or prochiral atoms in the main chain in a unique arrangement, with respect to their adjacent constitutional units. In a syndiotactic PP, the configurational repeating unit consists of two configurational base units that are enantiomeric [53].

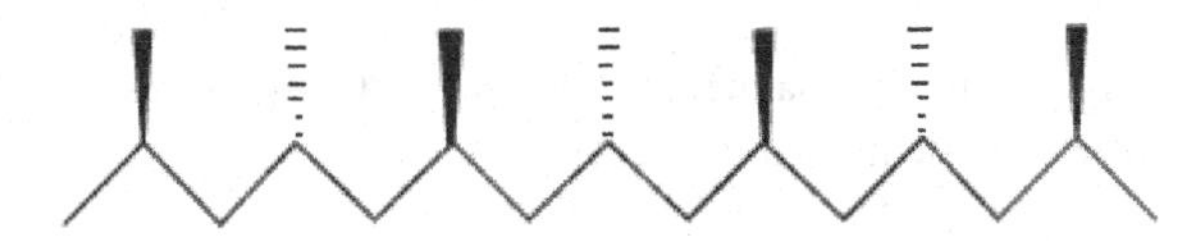

Pyrolysis of PP is based on free radical mechanism which begins with the homolytic breakage of the polymer chain (Fig. 4.5). The pyrolysis of PP has been studied at different levels of detail in the literature [53–55]. Overall, the major products from PP pyrolysis were determined to be as follows (in decreasing order): 2,4-dimethyl-1-heptene (propylene trimer), *n*-pentane, 2-methyl-1-pentene (propylene dimer), and propylene monomer [55].

The significant peaks of degradation of PP at 700°C (Fig. 4.6) by using the furnace pyrolyzer are propylene (t_R = 5.34 min) and branched C_6 and C_9 alkenes (oligomers), such as 2-methyl-1-pentene (propylene dimer, t_R = 6.34 min) and 2,4-dimethyl-1-heptene (propylene trimer, t_R = 11.88 min). Pyrolysis at lower temperatures (e.g. 550°C) leads to generate additionally of 2,4,6-trimethyl-1-nonene (propylene tetramer) and 2,4,6,8-tetramethyl-1-undecene (propylene pentamer) [55, 56]. The pyrolysis products of PP at 700°C are summarized in Table 4.1.

4.2.3. Characterization of poly(1-butene) (PB-1)

Poly(1-butene) [polybutylene, polybutene-1, PB-1] is a polyolefin of the chemical formula $(C_4H_8)_n$. It is produced by polymerization of 1-butene using supported Ziegler–Natta catalysts. PB-1 is a high-MW, linear, isotactic, and semi-crystalline polymer. This polymer combines typical characteristics

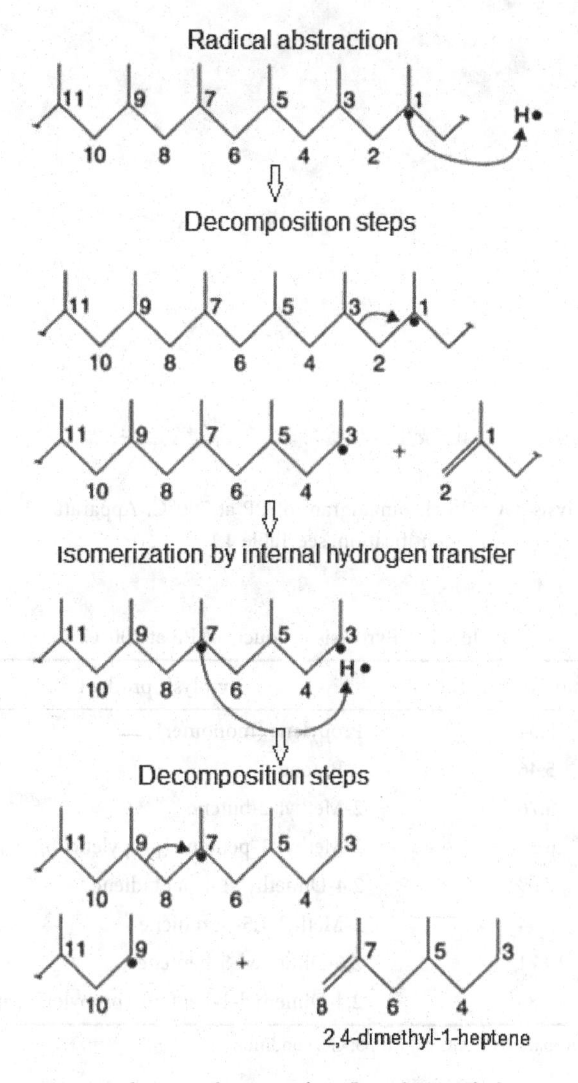

Fig. 4.5. Pyrolysis mechanism of PP (based on Refs. [53–55]).

of conventional polyolefins with certain properties of technical polymers [57]. PB-1, when applied as a pure or reinforced resin, can replace materials like metal, rubber, and engineering polymers. It is also used synergistically as a blend element to modify the characteristics of other polyolefins like PE or PP. PB-1 resins are used in customer applications such as easy-open

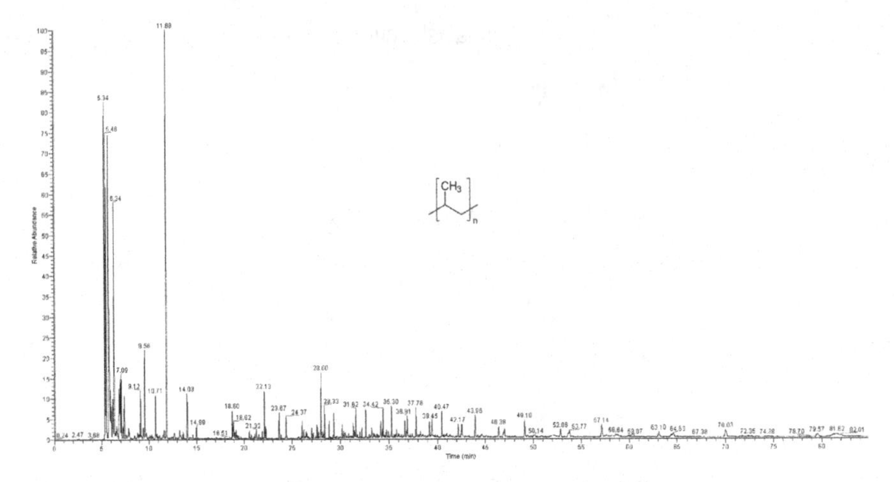

Fig. 4.6. Pyrolysis–GC/MS chromatogram of PP at 700°C. Apparatus 2, GC column 3, GC conditions 3. For peak identification, see Table 4.1.

Table 4.1. Pyrolysis products of PP at 700°C.

Retention time t_R (min)	Pyrolysis product
5.34	Propylene (monomer)
5.46	2-Butene
5.76	2-Methyl-2-butene
6.34	2-Methyl-1-pentene (propylene dimer)
7.09	2,4-Dimethyl-1,4-pentadiene
7.44	2-Methyl-1,5-hexadiene
10.71	2,6-Dimethyl-3-heptene
11.88	2,4-Dimethyl-1-heptene (propylene trimer)

Notes: Apparatus 2, GC column 3, GC conditions 3.

packaging, film modification, hot melt adhesives, polyolefin modification, hot water tanks, and pipe and fittings. They are light weight and have excellent performance characteristics, such as outstanding creep resistance and flexibility over a wide temperature range, low noise transmission, and good heat fusion properties [58]. Figure 4.7 shows the typical pyrolysis–GC/MS

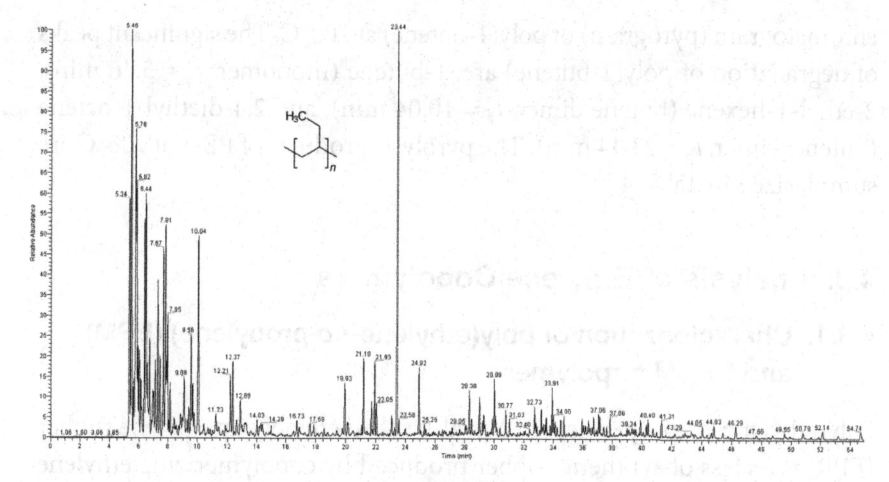

Fig. 4.7. Pyrolysis–GC/MS chromatogram of poly(1-butene) (PB-1) at 700°C. Apparatus 2, GC column 3, GC conditions 3. For peak identification, see Table 4.2.

Table 4.2. Pyrolysis products of poly(1-butene) (PB-1) at 700°C.

Retention time t_R (min)	Pyrolysis product
5.34	Propylene
5.46	1-Butene (monomer)
5.76	2-Methyl-1-butene
5.82	1,3-Pentadiene
6.35	1-Hexene
6.44	3-Methyl-1-pentene
6.72	1,3-Hexadiene
7.11	2,4-Hexadiene
7.28	2-Ethyl-1,3-butadiene
7.81	3-Methylhexane
7.95	2-Heptene
8.13	3-Heptene
10.04	2-Ethyl-1-hexene (butene dimer)
23.44	2,4-Diethyl-1-octene (butene trimer)

Notes: Apparatus 2, GC column 3, GC conditions 3.

chromatogram (pyrogram) of poly(1-butene) at 700°C. The significant peaks of degradation of poly(1-butene) are: 1-butene (monomer, $t_R = 5.46$ min), 2-ethyl-1-hexene (butene dimer, $t_R = 10.04$ min), and 2,4-diethyl-1-octene (butene trimer, $t_R = 23.44$ min). The pyrolysis products of PB-1 at 700°C are summarized in Table 4.2.

4.3. Analysis of Ethylene Copolymers

4.3.1. Characterization of poly(ethylene-*co*-propylene) (EPM) and EPDM terpolymer

Poly(ethylene-*co*-propylene) (EPM), also called ethylene–propylene rubber (EPR) is a class of synthetic rubber produced by copolymerizing ethylene and propylene, usually in combination with other chemical compounds [59]. In addition to elastic properties, ethylene–propylene copolymers display excellent resistance to electricity and ozone and an ability to be processed with a number of additives. They are made into products for use in automotive engines, electrical wiring, and construction. There are two major types of ethylene–propylene copolymers with elastic properties: those made with ethylene and propylene alone, and those made with small amounts (ca. 5%) of a diene, usually 5-ethylidene-2-norbornene (ENB), 1,4-hexadiene (HD) or dicyclopentadiene (DCPD) [60]. The former is known as ethylene–propylene monomer (EPM) and the latter as ethylene–propylene-diene monomer (EPDM). The copolymers contain approximately 60% (w/w) ethylene. The principal uses of EPM are in automobile parts and as an impact modifier for PP. In EPDM the ethylene content is around 45–75% (w/w). EPDM is employed in flexible seals for automobiles, wire and electrical insulation, weather stripping, tire sidewalls, hoses, and roofing film as well as in building and construction [59, 60]. Figure 4.8 shows a typical Py–GC/MS TIC chromatogram obtained by pyrolysis of commercially available poly(ethylene-*co*-propylene) (EPM) at 700°C. The pyrolysis products identified by using mass spectra library *NIST 05* are summarized in Table 4.3. Figure 4.9, on the other hand, shows the pyrolysis–GC/MS TIC chromatogram of EPDM having 5-ethylidene-2-norbornene (ENB) as a diene at 700°C. The pyrolysis products identified by using the same mass spectra library *NIST 05* are summarized in Table 4.4. As can be seen

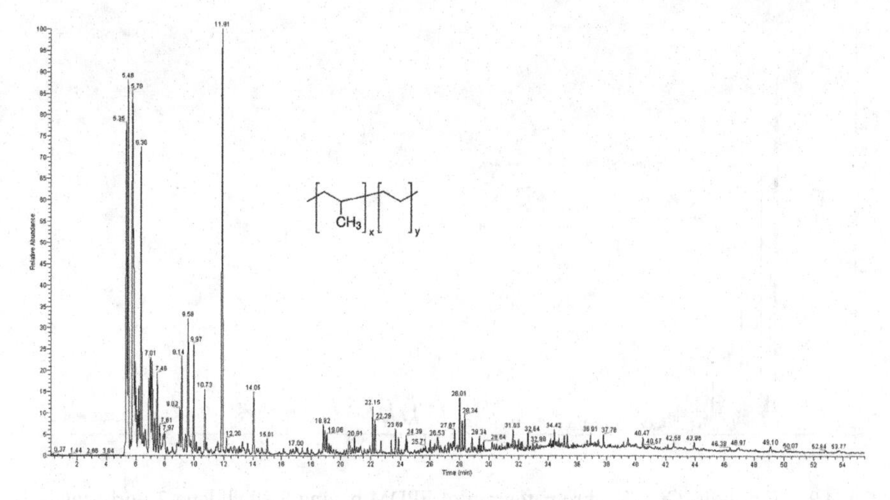

Fig. 4.8. Pyrolysis–GC/MS chromatogram of poly(ethylene-*co*-propylene) (EPM) at 700°C. Apparatus 2, GC column 3, GC conditions 3. For peak identification, see Table 4.3.

Table 4.3. Pyrolysis products of of poly(ethylene-*co*-propylene) (EPM) at 700°C.

Retention time t_R (min)	Pyrolysis product
5.35	Propylene
5.48	1-Butene
5.78	2-Methyl-1-butene
5.85	1,3-Pentadiene
6.19	2-Methyl-2-pentene
6.36	2-Methyl-1-pentene (propylene dimer)
7.12	2,4-Dimethyl-1,4-pentadiene
7.46	2-Methyl-1,5-hexadiene
9.03	2,4-Dimethyl-1-hexene
9.14	3,5-Dimethyl-2-hexene
9.58	2-Methyl-1,5-heptadiene
9.97	2-Methyl-1-heptene
10.73	2,6-Dimethyl-3-heptene
11.91	2,4-Dimethyl-1-heptene (propylene trimer)

Notes: Apparatus 2, GC column 3, GC conditions 3.

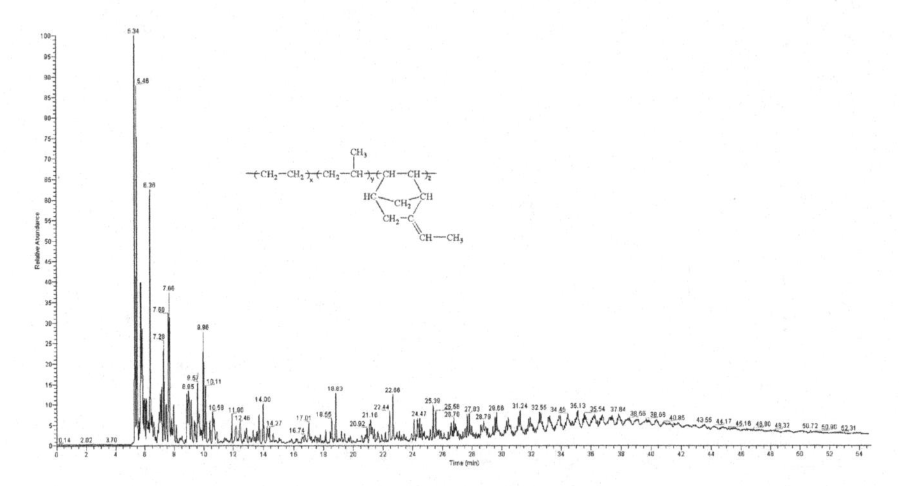

Fig. 4.9. Pyrolysis–GC/MS chromatogram of EPDM having 5-ethylidene-2-norbornene (ENB) as a diene at 700°C. Apparatus 2, GC column 3, GC conditions 3. For peak identification, see Table 4.4.

Table 4.4. Pyrolysis products of EPDM having 5-ethylidene-2-norbornene (ENB) as a diene at 700°C.

Retention time t_R (min)	Pyrolysis product
5.34	Propylene
5.46	2-Butene
5.62	3-Methyl-1-butene
5.74	2-Pentene
5.83	1,3-Pentadiene
6.36	2-Methyl-1-pentene (propylene dimer) + 1-hexene
7.06	3-Methyl-1-hexene
7.28	Benzene
7.60	2-Methyl-1-hexene
7.66	1-Heptene
7.82	*n*-Heptane
7.95	2-Heptene
9.57	Toluene
9.96	2-Methyl-1-heptene

(*Continued*)

Table 4.4. (*Continued*)

Retention time t_R (min)	Pyrolysis product
10.11	1-Octene
11.90	2,4-Dimethyl-1-heptene (propylene trimer)
12.88	Ethylbenzene
13.73	2-Methyl-1-octene
14.00	1-Nonene
14.27	Styrene
14.40	*n*-Nonane
14.65	4-Nonene
18.55	2-Methyl-1-nonene
18.83	1-Decene
22.44	2-Methyl-1-decene
22.66	1-Undecene
25.39	2-Methyl-1-undecene
25.56	1-Dodecene
27.68	2-Methyl-1-dodecene
27.83	1-Tridecene
29.55	2-Methyl-1-tridecene
29.66	1-Tetradecene
31.14	2-Methyl-1-tetradecene
31.24	1-Pentadecene

Notes: Apparatus 2, GC column 3, GC conditions 3.

from Figs. 4.8 and 4.9, propylene (t_R = 5.34–5.35 min) is a common pyrolysis product formed from both ethylene and propylene sequences in both EPM and EPDM copolymers. Propylene is the most abundant pyrolysis product of EPDM. The significant peaks of degradation of EPM (Fig. 4.8 and Table 4.3) are (except propylene) 2-methyl-1-pentene (propylene dimer) (t_R = 6.36 min) and 2,4-dimethyl-1-heptene (propylene trimer) (t_R = 11.91 min) from the propylene sequence of the copolymer. Among the pyrolysis products of EPDM (Fig. 4.9 and Table 4.4), the characteristic eluted pairs consist of 2-methyl-1-alkene and 1-alkene, in the order of increasing $n + 1$ carbon number in the molecule. They are formed during pyrolysis of the propylene sequence and the ethylene sequence of the terpolymer, respectively. According to the results of investigation of Choi and Kim [60], benzene ($t_R = 7.28$ min) and toluene ($t_R = 9.57$ min) are pyrolysis products of EPDM formed from ENB units. In my opinion, these aromatics, as well as the identified ethylbenzene ($t_R = 12.88$ min) and styrene ($t_R = 14.27$ min), could also have arisen in the pyrolysis of the EPDM additives.

4.3.2. Characterization of poly(ethylene-*co*-1-butene)

Poly(ethylene-*co*-1-butene) is a linear copolymer which exhibits plastic and elastic characteristics. As a low temperature impact modifier with high clarity, it reduces stress cracking and has a low temperature heat sealability. Poly(ethylene-*co*-1-butene) is used as an additive for PE and PP in injection, extrusion, and blow molding. Figure 4.10 shows the typical pyrolysis-GC/MS chromatogram (pyrogram) of poly(ethylene-*co*-1-butene) at 700°C. The pyrolysis products identified by using the mass spectra library *NIST 05* are summarized in Table 4.5.

4.3.3. Characterization of poly(ethylene-*co*-vinyl acetate)

Poly(ethylene-*co*-vinyl acetate) (EVA) is produced by copolymerization of ethylene and vinyl acetate (VA). With increasing proportion of the polar

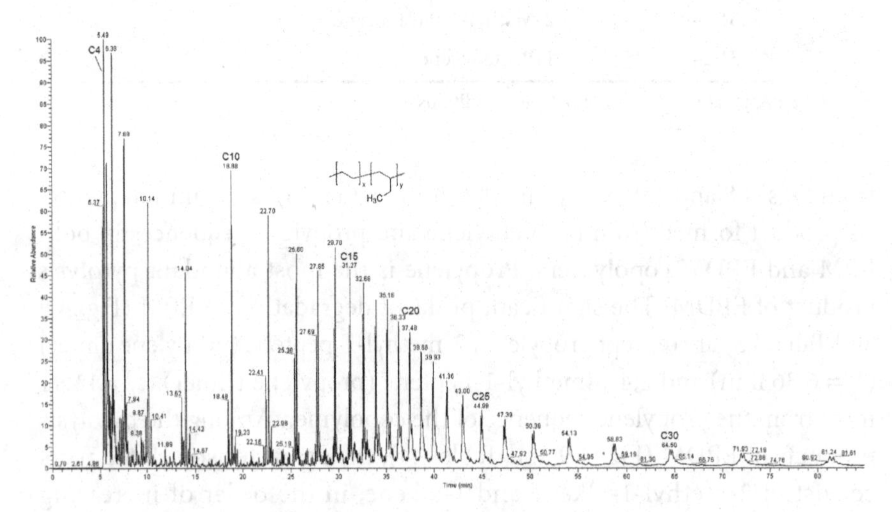

Fig. 4.10. Pyrolysis–GC/MS chromatogram of poly(ethylene-*co*-1-butene) at 700°C. Apparatus 2, GC column 3, GC conditions 3. For peak identification, see Table 4.5.

Table 4.5. Pyrolysis products of poly(ethylene-*co*-1-butene) at 700°C.

Retention time t_R (min)	Pyrolysis product
5.37	Propylene
5.49	2-Butene
5.76	2-Pentene
5.94	1,3-Pentadiene
6.38	1-Hexene
7.14	2,4-Hexadiene
7.45	5-Methyl-1,3-cyclopentadiene
7.69	1-Heptene
7.84	*n*-Heptane
9.87	1,7-Octadiene
10.41	2,4-Dimethylhexane
13.62	1,8-Nonadiene
13.81	2-Ethyl-1-hexene
14.04	1-Nonene
14.29	4-Nonene
14.42	*n*-Nonane
18.48	1,9-Decadiene
18.59	2-Ethyl-1-octene
18.88	1-Decene
19.23	*n*-Decane
22.41	1,10-Undecadiene
22.70	1-Undecene
22.96	*n*-Undecane
25.38	1,11-Dodecadiene
25,60	1-Dodecene
25.80	*n*-Dodecane
27.69	1,12-Tridecadiene
27.85	1-Tridecene
28.00	*n*-Tridecane
29.57	1,13-Tetradecadiene

(*Continued*)

Table 4.5. (*Continued*)

Retention time t_R (min)	Pyrolysis product
29.70	1-Tetradecene
29.81	*n*-Tetradecane
31.16	1,14-Pentadecadiene
31.27	1-Pentadecene
31.37	*n*-Pentadecane
32.59	1,15-Hexadecadiene
32.68	1-Hexadecene
32.76	*n*-Hexadecane
33.89	1,16-Heptadecadiene
33.97	1-Heptadecene
34.04	*n*-Heptadecane
35.09	1,17-Octadecadiene
35.16	1-Octadecene
35.23	*n*-Octadecane
36.25	1,18-Nonadecadiene
36.33	1-Nonadecene
36.38	*n*-Nonadecane
37.42	1,19-Eicosadiene
37.48	1-Eicosene
37.54	*n*-Eicosane
38.60	1,20-Heneicosadiene
38.66	1-Heneicosene
38.73	*n*-Heneicosane
39.87	1,21-Docosadiene
39.93	1-Docosene
40.00	*n*-Docosane
41.28	1,22-Tricosadiene
41.36	1-Tricosene
41.42	*n*-Tricosane
42.92	1,23-Tetracosadiene
43.00	1-Tetracosene

(*Continued*)

Table 4.5. (*Continued*)

Retention time t_R (min)	Pyrolysis product
43.07	*n*-Tetracosane
44.88	1,24-Pentacosadiene
44.99	1-Pentacosene
45.06	*n*-Pentacosane
47.28	1,25-Hexacosadiene
47.39	1-Hexacosene
47.48	*n*-Hexacosane
50.24	1,26-Heptacosadiene
50.39	1-Heptacosene
50.50	*n*-Heptacosane
53.97	1,27-Octacosadiene
54.13	1-Octacosene
54.27	*n*-Octacosane
58.64	1,28-Nonacosadiene
58.83	1-Nonacosene
59.00	*n*-Nonacosane
64.50	1,29-Triacontadiene
64.76	1-Triacontene
64.92	*n*-Triacontane
71.93	1,30-Hentriacontadiene
72.19	1-Hentriacontene
72.46	*n*-Hentriacontane
81.24	1,31-Dotriacontadiene
81.61	1-Dotriacontene
81.92	*n*-Dotriacontane

Notes: Apparatus 2, GC column 3, GC conditions 3.

co-monomer VA, the products change from modified PE to rubber-like products [61]. EVA is mainly recognized for its flexibility and toughness (even at low temperatures), adhesion characteristics, and stress-cracking resistance. Compared to LDPE, EVA is more polar and less crystalline due to the acetate groups. With increasing VA content, EVA copolymer becomes softer due to the decreased crystallinity. Up to a VA content of 10%, the density decreases and the crystalline structure is not destroyed [61]. While higher densities usually mean higher stiffness and a higher glass transition temperature, the opposite is true in the case of EVA copolymers. Transparency increases with increasing VA content. Products with up to 10% of VA are more transparent, flexible, and tougher than LDPE [61]. The high resistance to ESC (Environmental stress cracking) is especially useful for cable isolation. Between 15% and 30% of VA, the products are comparable to plasticized PVC. They are very soft and flexible. Compounds with 30–40% of VA are soft, elastic, and highly fillable. Strength and adhesion are the desirable properties for coatings and adhesives. Between 40% and 50% of VA rubber-like properties predominate and these products can be crosslinked as cable insulation by either peroxide or radiation. Copolymers with 70–95% of VA are used for manufacturing of emulsion paints, adhesives, and film coatings [61]. EVA is resistant to dilute mineral acids, alkaline substances, alcohols, fats, oils, and detergents but not to concentrated mineral acids, ketones, and aromatic or chlorinated hydrocarbons. The resistance to ESC increases with increasing VA content and a decreasing melt index. It is significantly higher for EVA copolymers than for comparable LDPE. EVA is one of the materials popularly known as *expanded rubber* or *foam rubber*. EVA foam is used as padding in equipment for various sports such as ski boots, bicycle saddles, hockey pads, boxing and mixed martial arts gloves and helmets, wakeboard boots, waterski boots, fishing rods, and fishing reel handles. It is typically used as a shock absorber in sports shoes. EVA is also used in biomedical engineering applications as a drug delivery device [62]. Figure 4.11 shows a typical Py–GC/MS TIC chromatogram obtained by pyrolysis of commercially available EVA at 700°C by the use of the furnace pyrolyzer. The pyrogram of this copolymer is almost identical with the pyrogram of PE except for a peak belonging to acetic acid (Fig. 4.12, $t_R = 5.97$ min).

4.3.4. Characterization of poly(ethylene-*co*-acrylic acid)

Random copolymers of ethylene and acrylic acid (pEAA) are industrially important materials that have found use in a wide range of applications [63]. The copolymers are used as a foil adhesive layer in juice and condiment packaging, as a heat sealant for tea bags and spice oil sachets, and as

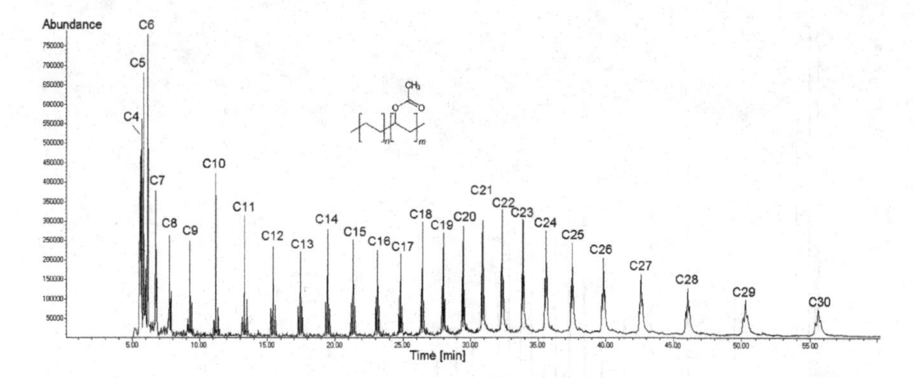

Fig. 4.11. Pyrolysis–GC/MS chromatogram of EVA at 700°C. Apparatus 1, GC column 1, GC conditions 2.

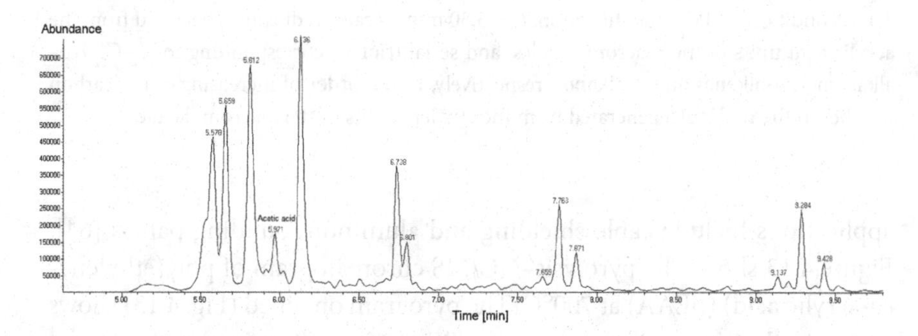

Fig. 4.12. Pyrolysis–GC/MS chromatogram of EVA at 700°C. Range 4.6–9.8 min. Apparatus 1, GC column 1, GC conditions 2. Peak identification: t_R = 5.97 min — acetic acid.

an adhesive in other aluminum applications [64]. Ethylene/acrylic acid and methacrylic acid (MAA) copolymer resins provide excellent adhesion to aluminum and other polar substrates, as well as to nylon and paper. As a seal layer in laminated and coextruded structures, the copolymer resin resists delamination and seal failure, even in aggressive chemical environments [64]. The benefits include excellent hot tack strength, toughness, and low-temperature seal initiation. It extends product shelf life and provides excellent seal-through contamination. It also allows high-speed packaging and results in fewer package failures, which help to reduce waste. Other

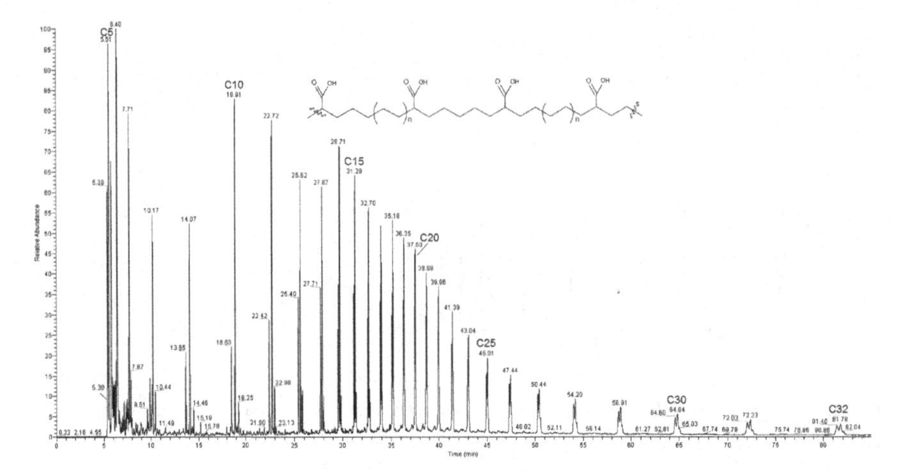

Fig. 4.13. Pyrolysis–GC/MS chromatogram of pEAA at 700°C. Apparatus 2, GC column 3, GC conditions 3. Peak identification: t_R = 5.30 min — carbon dioxide, generated from the acrylic acid units of the macromolecules, and serial triplets, corresponding to C_3–C_{32} α,ω-alkadienes, α-alkenes and *n*-alkanes, respectively, in the order of increasing $n + 1$ carbon number in the molecule, generated from the ethylene units of the macromolecules.

applications include cable shielding and aluminum building panels [64]. Figure 4.13 shows the pyrolysis–GC/MS chromatogram of poly(ethylene-*co*-acrylic acid) (pEAA) at 700°C. The pyrogram obtained (Fig. 4.13) shows many similarities to the pyrogram of PE at 700°C (see Section 4.2.1) and also consists of serial triplets, corresponding to C_3–C_{32} α,ω-alkadienes, α-alkenes, and *n*-alkanes, respectively, in the order of increasing $n + 1$ carbon number in the molecule, like in PE pyrolysis. The compounds are generated from the ethylene units of the macromolecules. The substance elute at the retention time t_R = 5.30 min corresponds to carbon dioxide, generated from the acrylic acid units of the macromolecules.

4.3.5. Characterization of a blend of a LLDPE (α-olefin copolymer) with poly(ethylene-*co*-butyl acrylate)

Poly(ethylene-*co*-butyl acrylate), often abbreviated EBA, is made by copolymerization under high hydrostatic pressure in the same reactors that

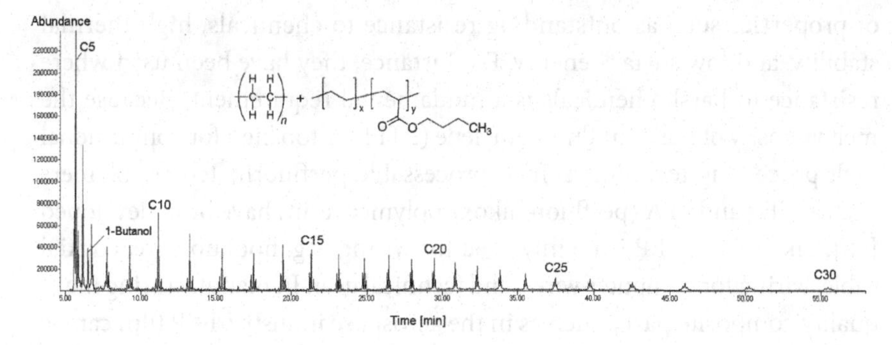

Fig. 4.14. Pyrolysis–GC/MS chromatogram of a blend of an LLDPE (α-olefin copolymer) with EBA at 700°C. Apparatus 1, GC column 1, GC conditions 2. Peak identification: t_R = 6.51 min — 1-butanol, generated from the butyl acrylate units of the macromolecules, and serial triplets, corresponding to C_3–C_{30} α,ω-alkadienes, α-alkenes and n-alkanes, respectively, in the order of increasing n+1 carbon number in the molecule, generated from the ethylene/PE parts of the macromolecules.

are used to make LDPE. The acrylic acid units reduce the crystallinity and make the polymer more polar than LDPE. EBA is currently used as semiconductive material in electrical insulation systems. The fact that EBA is manufactured by the high-pressure polymerization process results in a polymer with short-chain and long-chain branching [65, 66]. Blends of an LLDPE with EBA are used for the sealing of roofs and nonwatertight buildings [67]. Figure 4.14 shows the pyrolysis–GC/MS chromatogram of a blend of an LLDPE (α-olefin copolymer) with EBA at 700°C. The identified pyrolysis products were 1-butanol (t_R = 6.51 min), generated from the butyl acrylate units of the macromolecules and serial triplets, corresponding to C_3–C_{30} α,ω-alkadienes, α-alkenes, and n-alkanes, respectively, in the order of increasing n + 1 carbon number in the molecule, generated from the ethylene/PE parts of the macromolecules.

4.3.6. Characterization of poly(tetrafluoroethylene-*co*-hexafluoropropylene) (FEP)

Fluorinated ethylene-propylene (FEP) is a random copolymer of hexafluoropropylene (HFP) and tetrafluoroethylene (TFE), produced by free-radical polymerization. Perfluorinated polymers possess a unique combination of properties such as outstanding resistance to chemicals, high thermal stability, and low surface energy. For instance, they have been used where resistance to harsh chemicals is a fundamental requirement. Because the melt viscosity of polytetrafluoroethylene (PTFE) is too high for conventional melt processing techniques, melt processable perfluorinated copolymers such as FEP and PFA (perfluoroalkoxy polymer resin) have been developed [68]. Like PTFE, FEP is mainly used for wiring, e.g. hookup wire, coaxial cable, wiring for computer wires, and technical gear. In manufacturing high-quality composite parts, such as in the aerospace industry, FEP film can be used to protect molds during the curing process. In such applications, the film is called "release film" and is intended to prevent the curing adhesive polymer (e.g. the epoxy in a carbon fiber/epoxy composite laminate) from bonding to the metal tooling. Being able to maintain chemical composure in extreme temperatures and resist damage from chemical fuels further makes FEP a suitable choice in the industry [69]. Due to its flexibility, extreme resistance to chemical attack, and optical transparency, this material, along with PFA is routinely used for plastic lab ware and tubing that involve critical or highly corrosive processes.

Figure 4.15 shows the pyrolysis–GC/MS chromatogram in full-scan mode of poly(tetrafluoroethylene-*co*-hexafluoropropylene) (FEP) at 700°C. Both pyrolysis products of FEP, the monomer TFE, generated from TFE part of the macromolecules, and the monomer HFP, generated from HFP part of the macromolecules elute together as one peak at 9.29 min.

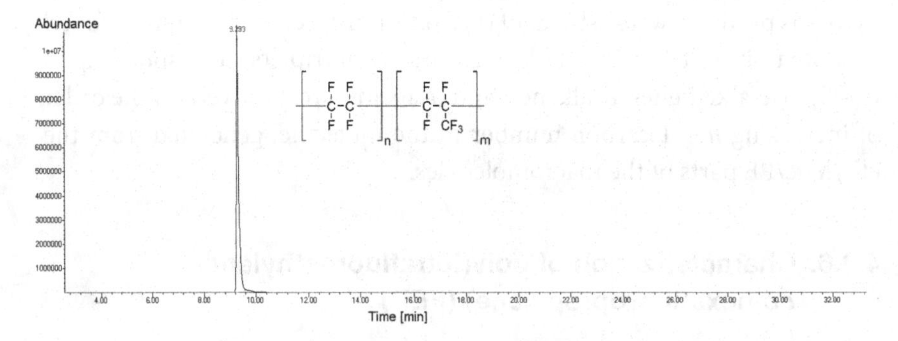

Fig. 4.15. Pyrolysis–GC/MS chromatogram in full-scan mode of poly(tetrafluoroethylene-*co*-hexafluoropropylene) (FEP) at 700°C. Apparatus 1, GC column 2, GC conditions 1. Peak identification: $t_R = 9.29$ min — TFE + HFP.

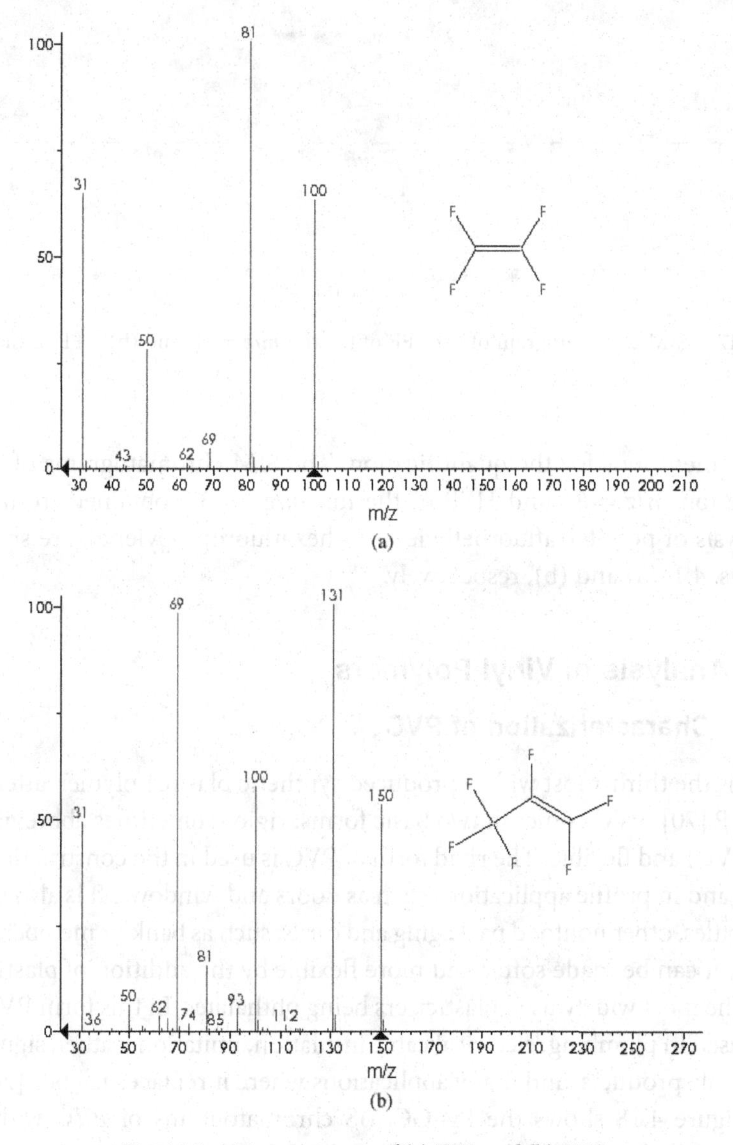

Fig. 4.16. Mass spectrum of (a) TFE, and (b) HFP.

To follow the appearance of the ion current curves of the base peak ions $m/z = 81$ of TFE and $m/z = 131$ of HFP (Figs. 4.16(a) and (b), respectively), the selected ion monitoring (SIM) mode was used. SIM mode in MS can be used for the detection of substances with known m/z of the molecular

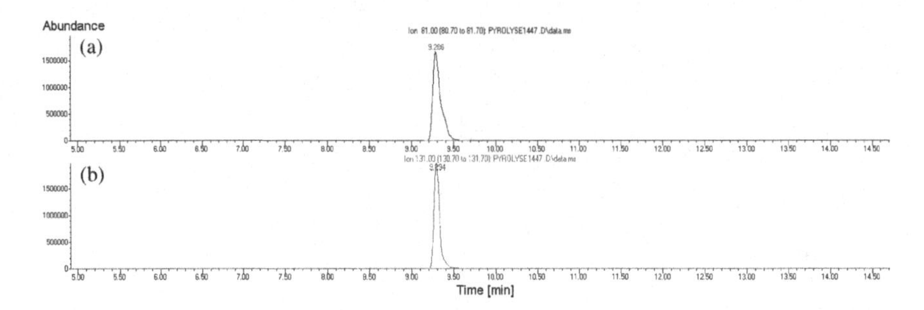

Fig. 4.17. SIM chromatogram of (a) TFE of the ion $m/z = 81$, and (b) HFP of the ion $m/z = 131$.

or fragment ions for the quantification. The SIM chromatograms of TFE of the ion $m/z = 81$ and HFP of the ion $m/z = 131$, obtained from the pyrolysis of poly(tetrafluoroethylene-*co*-hexafluoropropylene), are shown in Figs. 4.17(a) and (b), respectively.

4.4. Analysis of Vinyl Polymers

4.4.1. Characterization of PVC

PVC is the third-most widely produced synthetic plastic polymer, after PE and PP [70]. PVC comes in two basic forms: rigid (sometimes abbreviated as RPVC) and flexible. The rigid form of PVC is used in the construction of pipes and in profile applications such as doors and windows. It is also used for bottles, other nonfood packaging and cards, such as bank or membership cards. It can be made softer and more flexible by the addition of plasticizers. The most widely used plasticizers being phthalates. In this form PVC is also used in plumbing, electrical cable insulation, imitation leather, signage, inflatable products, and many applications where it replaces rubber [70].

Figure 4.18 shows the Py–GC/MS chromatograms of PVC without additives and PVC with bis(2-ethylhexyl) phthalate (BEHP) plasticizer at 700°C. The identification results of pyrolysis products of both PVC without additives and PVC with BEHP plasticizer are summarized in Table 4.6. As can be seen from Fig. 4.18(a), the main feature of the pyrolysis of PVC at 700°C is the formation of hydrogen chloride (HCl) ($t_R = 5.40$ min) and the formation of aromatic hydrocarbons. Benzene ($t_R = 6.48$ min) is the

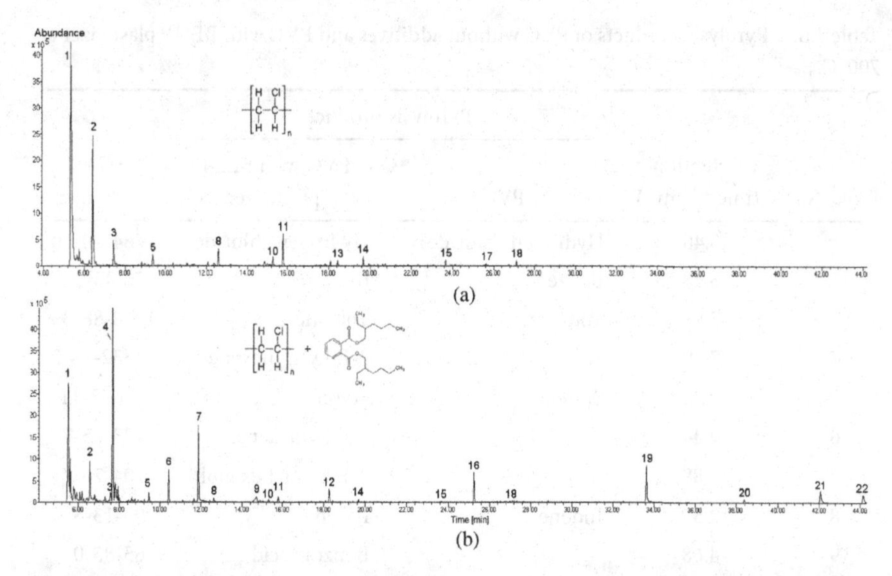

Fig. 4.18. Pyrolysis–GC/MS chromatogram of (a) PVC without additives, and (b) PVC with BEHP plasticizer at 700°C. Apparatus 1, GC column 1, GC conditions 2. For peak identification, see Table 4.6.

major pyrolysis product of PVC from the aromatic hydrocarbons. Other aromatic components, like toluene (t_R = 7.50 min) and the polycyclic aromatic hydrocarbons (PAH) are also present (Fig. 4.18, Table 4.6). This is the result of the formation of double bonds by the elimination of HCl from the PVC macromolecules, followed by the breaking of the carbon chain with or without cyclization [30, 71, 72]. The side-group scission occurs when the side groups attached to the backbone are broken away, resulting in the backbone becoming polyunsaturated. A two-step degradation mechanism begins with the elimination of HCl from the polymer chain leaving a polyunsaturated backbone that, upon further heating, produces the characteristic aromatics (Fig. 4.19).

The formation of aromatic compounds in PVC pyrolysis is schematically exemplified for benzene in the monograph of Moldoveanu [30]. The results of pyrolysis of PVC with BEHP plasticizer at 600°C was described in earlier work of the author (P. K.) [12]. The thermal decomposition of BEHP plasticizer at 700°C (Fig. 4.18(b) and Table 4.6) leads to the formation of

Table 4.6. Pyrolysis products of PVC without additives and PVC with BEHP plasticizer at 700°C.

		Pyrolysis product		
Peak No.	Retention time t_R (min)	PVC	PVC with BEHP plasticizer	CAS number
1	5.40	Hydrogen chloride	Hydrogen chloride	7647-01-0
2	6.48	Benzene	Benzene	71-43-2
3	7.50	Toluene	Toluene	108-88-3
4	7.69		2-Ethyl-1-hexene	1632-16-2
5	9.40	Styrene	Styrene	79637-11-9
6	10.44		2-Ethylhexanal	123-05-7
7	11.88		2-Ethyl-1-hexanol	104-76-7
8	12.59	Indene	Indene	95-13-6
9	14.68		Benzoic acid	65-85-0
10	15.28	Methylindene	Methylindene	29036-25-7
11	15.80	Naphthalene	Naphthalene	91-20-3
12	18.26		Phthalic anhydride	85-44-9
13	18.43	2-Methylnaphthalene		91-57-6
14	19.67	Biphenyl	Biphenyl	92-52-4
15	23.67	Fluorene	Fluorene	86-73-7
16	25.25		2-Ethylhexyl benzoate	5444-75-7
17	25.62	1,2-Diphenylethylene		103-30-0
18	27.12	Anthracene	Anthracene	120-12-7
19	33.64		Not identified	
20	38.39		BEHP	117-81-7
21	42.07		Bis(2-ethylhexyl) isophthalate	137-89-3
22	44.15		Diisooctyl phthalate	27554-26-3

Notes: Apparatus 1, GC column 1, GC conditions 2.

2-ethyl-1-hexene (t_R = 7.69 min), 2-ethylhexanal (t_R = 10.44 min), 2-ethyl-1-hexanol (t_R = 11.88 min), benzoic acid (t_R = 14.68 min), phthalic anhydride (t_R = 18.26 min), and 2-ethylhexyl benzoate (t_R = 25.25 min). The mass spectra of the significant pyrolysis products of BEHP plasticizer are

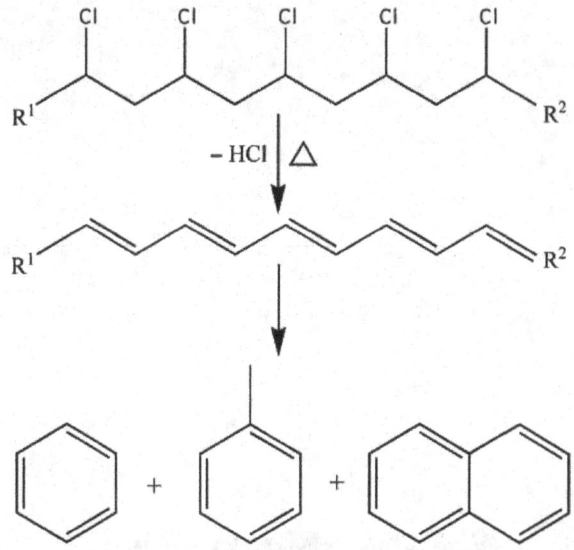

Fig. 4.19. Side-group scission reaction mechanism of thermal degradation of PVC [71, 72].

shown in Fig. 4.20. The identification results are consistent with the previously published works of Bove and Dalven [73] and Saido *et al.* [74].

4.4.2. Characterization of poly(vinyl alcohol)

Poly(vinyl alcohol) (PVA) was first prepared by Hermann and Haehnel in 1924 by hydrolyzing poly(vinyl acetate) in ethanol with potassium hydroxide [75]. PVA is produced commercially from poly(vinyl acetate), usually by a continuous process. The acetate groups are hydrolyzed by ester interchange with methanol in the presence of anhydrous sodium methylate or aqueous sodium hydroxide. PVA is an odorless and tasteless, translucent, white- or cream-colored granular powder. It is soluble in water, slightly soluble in ethanol, but insoluble in other organic solvents. PVA is divided into two classes — partially hydrolyzed PVA and fully hydrolyzed PVA. Partially hydrolyzed PVA is used in food as a moisture barrier film, for food supplement tablets, and for dry food with inclusions that need to be protected from moisture uptake. PVA has various applications in the food industries as a binding and coating agent. PVA protects the active ingredients from

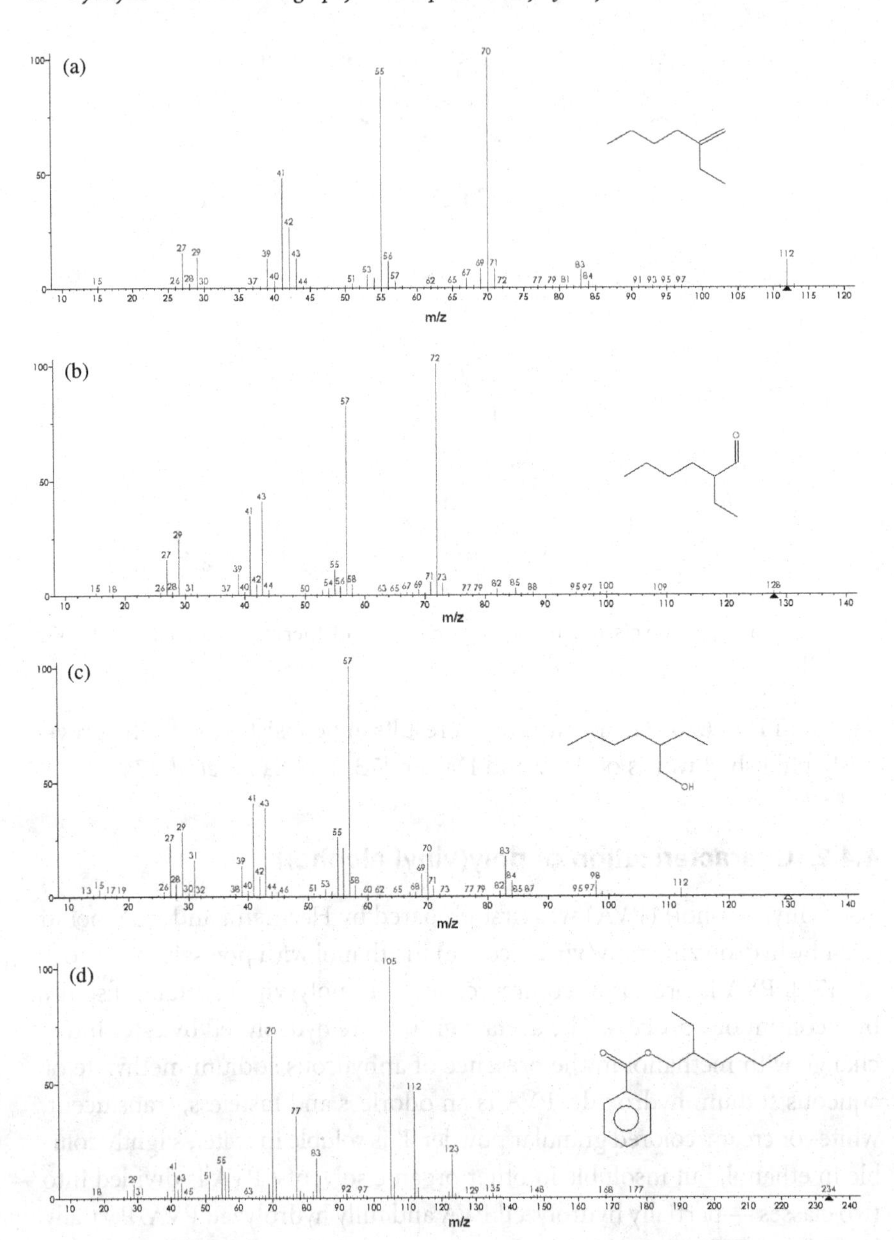

Fig. 4.20. Mass spectra of the significant degradation products of BEHP plasticizer: (a) 2-ethyl-1-hexene, (b) 2-ethylhexanal, (c) 2-ethyl-1-hexanol, (d) 2-ethylhexyl benzoate.

moisture, oxygen, and other environmental components while simultaneously masking their taste and odor. PVA may be used in high-moisture foods in order to retain the overall satisfactory taste, texture, and quality of the foods. Confectionery products may also contain PVA in order to preserve the integrity of the moisture-sensitive constituents. PVA is used as a water-soluble film useful for packaging. An example is the envelope containing laundry detergent in "liqui-tabs." This polymer is used also as packaging material for dishwasher tabs.

PVA hydrogels are three-dimensional (3D) network polymers that are physically or chemically crosslinked and are able to absorb large amounts of water [76]. Due to its higher water content, microporous structure, and favorable mechanical and lubricating properties, the PVA hydrogels have been extensively recognized as a potential material for a variety of applications in the pharmaceutical, biomedical industry, especially in artificial biomaterials for cartilage repair [76].

Figure 4.21 shows the pyrolysis–GC/MS chromatogram of PVA at 700°C. The pyrolysis products identified by using the *NIST 05* mass spectra library are summarized in Table 4.7.

4.4.3. Characterization of poly(vinyl pyrrolidone)

Poly(vinyl pyrrolidone) (PVP) is a highly polar, amphoteric polymer, which is usually synthesized by free-radical polymerization of *N*-vinyl-2-pyrrolidone

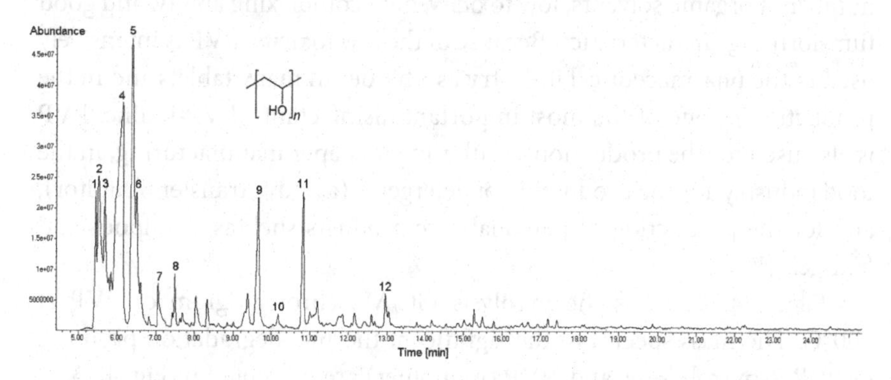

Fig. 4.21. Pyrolysis–GC/MS chromatogram of PVA at 700°C. Apparatus 1, GC column 1, GC conditions 2. For peak identification, see Table 4.7.

Table 4.7. Pyrolysis products of of PVA at 700°C.

Peak No.	Retention time t_R (min)	Pyrolysis product
1	5.45	Propene
2	5.52	Acetaldehyde
3	5.69	Acetone
4	6.14	Acetic acid
5	6.39	2,3-Dihydrofuran
6	6.50	Benzene
7	7.06	3-Penten-2-one
8	7.50	Toluene
9	9.66	2,4-Hexandienal
10	10.18	2-Cyclohexen-1-one
11	10.82	Benzaldehyde
12	12.97	Acetophenone

Notes: Apparatus 1, GC column 1, GC conditions 2.

(NVP) in water or alcohol, with a suitable initiator [77]. By selecting polymerization conditions, a wide range of MWs can be obtained, extending from low MW values (of a few thousand g/mol) to very high ones (2.0×10^6 g/mol or more). PVP is widely used in a variety of industries because of its unique properties, particularly good solubility not only in water but also in a large number of organic solvents, low toxicity, high complexing ability, and good film-forming characteristics. Because of the low toxicity, PVP is intensively used in the pharmaceutical industry as a binder in many tablets and in the production of one of the most important disinfectants, PVP-iodine. PVP is also used for the production of adhesives in paper manufacturing, in the food industry, for the production of detergents (as a dye transfer inhibitor), and for the production of personal care products such as shampoos and hair gels [77].

Figure 4.22 shows the pyrolysis–GC/MS chromatogram of PVP at 700°C. The mass spectra of the significant thermal degradation products of PVP, 2-pyrrolidone, and NVP (monomer) are presented in Fig. 4.23.

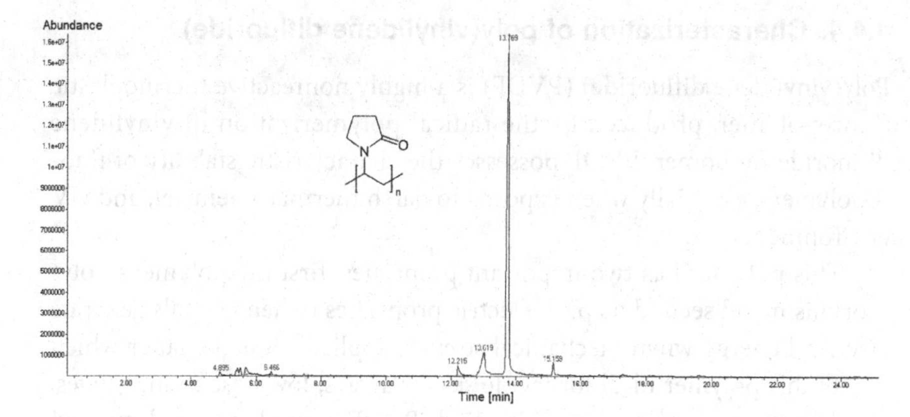

Fig. 4.22. Pyrolysis–GC/MS chromatogram of PVP at 700°C. Apparatus 1, GC column 1, GC conditions 2. Peak identification: t_R = 13.02 min — 2-pyrrolidone, t_R = 13.77 min — NVP (monomer).

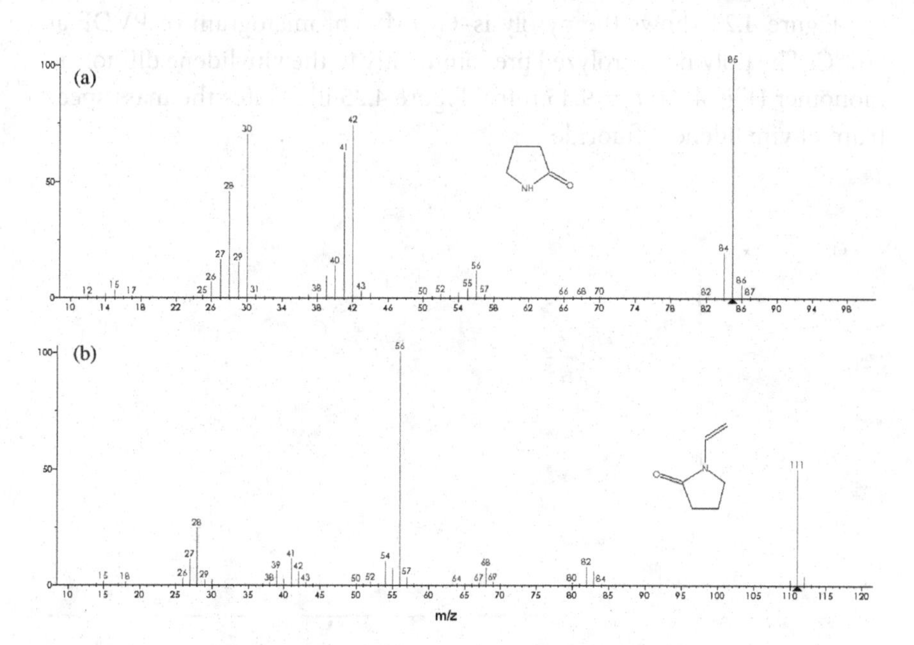

Fig. 4.23. Mass spectra of the significant pyrolysis products of PVP: (a) 2-pyrrolidone, (b) NVP (monomer).

4.4.4. Characterization of poly(vinylidene difluoride)

Poly(vinylidene difluoride) (PVDF) is a highly nonreactive thermoplastic fluoropolymer, produced by the radical polymerization of vinylidene difluoride monomer. PVDF possesses the characteristic stability of fluoropolymers, especially when exposed to harsh thermal, chemical, and UV environments.

This polymer has two important properties: first the polymer's polymorphism and second its piezoelectric properties (when crystals generate electrical energy when mechanical stress is applied). It is the latter which makes this polymer ideal for tactile sensor arrays, low-cost strain gauges, and lightweight audio transducers [78]. PVDF is used in a multitude of applications across aerospace, biotechnology, and electronic industries (e.g. robotics, sensors, and electrical wire insulation). It is also used to make hollow fibers and flat sheet tubular membranes for the medical and food beverage industry [78].

Figure 4.24 shows the pyrolysis–GC/MS chromatogram of PVDF at 700°C. The polymer pyrolyzed predominantly to the vinylidene difluoride monomer (Fig. 4.24, t_R = 9.45 min). Figure 4.25 illustrates the mass spectrum of vinylidene difluoride.

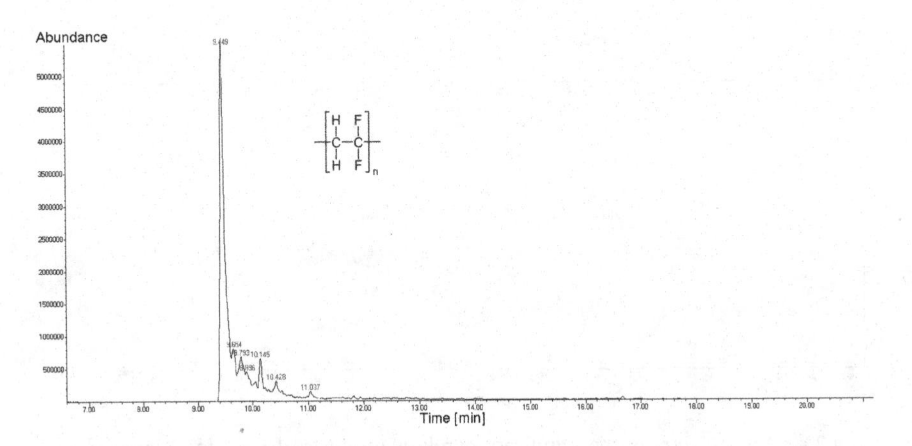

Fig. 4.24. Pyrolysis–GC/MS chromatogram of PVDF at 700°C. Apparatus 1, GC column 1, GC conditions 1. Peak identification: t_R = 9.45 min — vinylidene difluoride (monomer) and silicon tetrafluoride, t_R = 10.14 min — 1,3,5-trifluorobenzene.

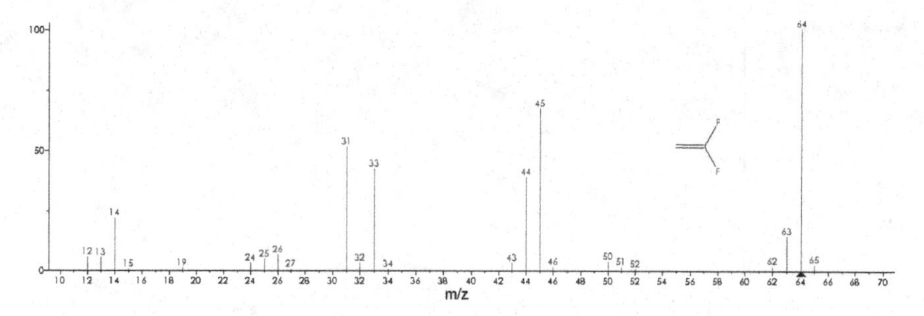

Fig. 4.25. Mass spectrum of vinylidene difluoride monomer.

4.5. Analysis of PSs

4.5.1. Characterization of PS

PSs form another class of polyhydrocarbons. The most common polymer in the class is PS, which is obtained by the polymerization of styrene, usually in the presence of peroxide initiator [30]. The styrene monomer is widely used in the industry as a starting material for PS, automobile rubber tires, glass-reinforced plastics, carpet coatings, and packaging material. PS has excellent characteristics necessary for the use in commercial and industrial products, either as pure material or as copolymers. These characteristics include the easy processing under injecting-molding, relatively good mechanical properties, transparency, good electrical insulating characteristics, etc. [30].

PS is solid in three main forms: crystalline or general purpose polymer, high-impact polymer and expanded polymer [1]. Commercial PS with an average MW typically in the range 50,000–200,000 g/mol is a hard and relatively brittle polymer. High-impact PS is produced by the addition of dienes. Expanded polystyrene (EPS) is produced by polymerizing styrene monomer and adding *iso*-pentane as a blowing agent. EPS is used for food packaging, for protection of products against damage during transport and storage, and in the building industry for insulation of exterior walls and foundations. PS is the second most widely used polymer in packaging [79].

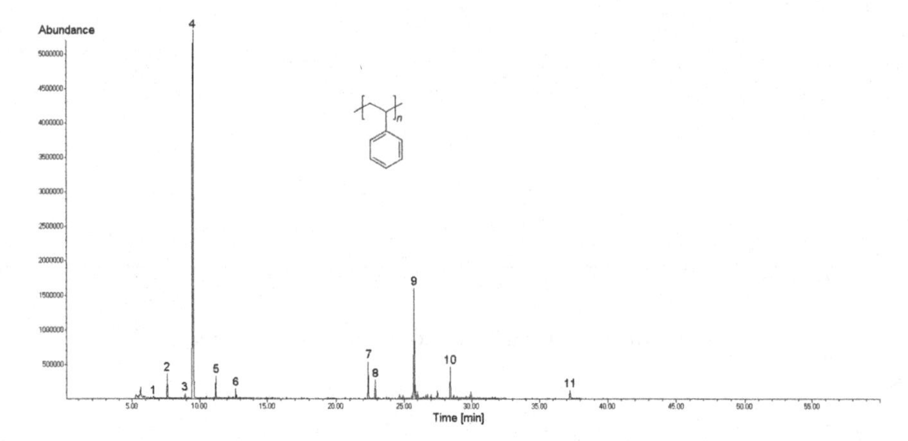

Fig. 4.26. Pyrolysis–GC/MS chromatogram of PS at 700°C. Apparatus 1, GC column 1, GC conditions 2. For peak identification, see Table 4.8.

Thermal decomposition of PS produces large amounts (ca. 85% m/m) of styrene, including the dimer/trimer of styrene and other products [80]. Literature data indicate that the styrene yield increases with temperature, while amounts of dimer and trimer decrease. The optimum conditions for the production of styrene by flash pyrolysis of PS were described by Bouster *et al.* [80]. The reaction of oligomer formation was explained. A typical pyrogram obtained by pyrolysis of PS at 700°C using the furnace pyrolyzer is given in Fig. 4.26. The degradation products identified by using GC/MS and *NIST 05* mass spectra library are summarized in Table 4.8. The mass spectrum of styrene monomer is presented in Fig. 4.80.

4.5.2. Analysis of styrene copolymers

The most popular styrene copolymers are produced by copolymerization of styrene with 1,3-butadiene (SBR), with acrylonitrile (SAN), with methyl methacrylate (SMMA), with 1,3-butadiene and acrylonitrile (ABS), with 1,3-butadiene, acrylonitrile and α-methylstyrene (ABS-α-MS), or with divinylbenzene (SDVB). The pyrolysis of SBR is described in Section 4.6.6, regarding rubber products. The pyrolysis of the other styrene copolymers is described in the following section.

Table 4.8. Pyrolysis products of PS at 700°C.

Peak No.	Retention time t_R (min)	Pyrolysis product
1	6.55	Benzene
2	7.56	Toluene
3	8.92	Ethylbenzene
4	9.58	Styrene
5	11.16	α-Methylstyrene
6	12.65	Indene
7	22.37	1,2-Diphenylethane
8	22.86	1,2-Diphenylpropane
9	25.70	Styrene dimer*
10	29.40	2,5-Diphenyl-1,5-hexadiene
11	37.16	Styrene trimer*

Notes: Apparatus 1, GC column 1, GC conditions 2.
*Identification based on literature data [56].

4.5.2.1. *Characterization of poly(styrene-co-acrylonitrile)*

Poly(styrene-*co*-acrylonitrile) (SAN) results from copolymerizing styrene with 20–30% acrylonitrile *via* a bulk or suspension route [1]. The resulting material has a higher softening point and higher impact strength than PS homopolymer but, while still transparent, it does carry a faint yellow pigmentation making it unsuitable for some household/food packaging applications [1]. In recent years, styrene–acrylonitrile copolymer (SAN) has successfully defended its application areas in the household sector, in cosmetic packaging, and in durable industrial batteries. It has done so by virtue of a special property profile comprised of chemical resistance, very good transparency, and high rigidity [79]. SAN finds also application in appliance knobs, refrigerator compartments, and syringes [1].

The main degradation products of SAN at 700°C are the monomers acrylonitrile and styrene (Fig. 4.27). The elimination of hydrogen cyanide (HCN) from the side chain of the copolymer (see the chemical structure in Fig. 4.27) generates double bonds in the backbone, which accelerates further decomposition due to the cleavage of the backbone in the ß-position

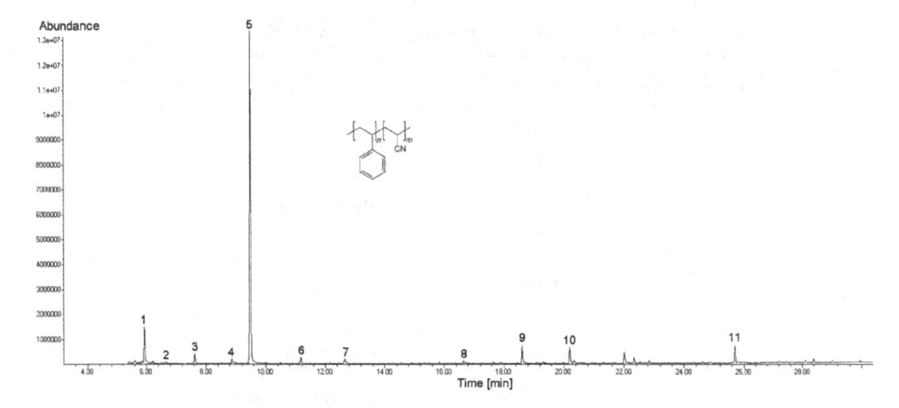

Fig. 4.27. Pyrolysis–GC/MS chromatogram of SAN at 700°C. Apparatus 1, GC column 1, GC conditions 2. For peak identification, see Table 4.9.

Table 4.9. Pyrolysis products of SAN at 700°C.

Peak No.	Retention time t_R (min)	Pyrolysis product
1	5.93	Acrylonitrile
2	6.64	Benzene
3	7.62	Toluene
4	8.90	Ethylbenzene
5	9.58	Styrene
6	11.16	α-Methylstyrene
7	12.65	Indene
8	16.67	Benzenepropanenitrile
9	18.61	2-Methylene-4-phenylbutanenitrile*
10	20.20	4-Phenylpent-4-enenitrile*
11	25.70	Styrene dimer*

Notes: Apparatus 1, GC column 1, GC conditions 2.
*Identification based on literature data [56].

to the double bond [30]. Both acrylonitrile monomer (Fig. 4.27, t_R = 5.93 min) and styrene monomer (Fig. 4.27, t_R = 9.58 min) are formed in the pyrolyzate in relatively large amounts, allowing an immediate identification of this copolymer as SAN. Other thermal decomposition products of SAN at 700°C are summarized in Table 4.9.

4.5.2.2. *Characterization of poly(acrylonitrile-co-1,3-butadiene-co-styrene)*

Poly(acrylonitrile-*co*-1,3-butadiene-*co*-styrene) (ABS) is made from acrylonitrile, 1,3-butadiene, and styrene by emulsion polymerization, bulk polymerization, or combined processes [1]. ABS is highly versatile and offers important property advantages over other materials in a wide variety of applications and industries. The material is also found in those applications and industries which are boosted by strong, lasting trends, such as demographic change, urbanization, and mobility [79]. ABS is widely used as a construction material in the automotive industry for vehicle fascia panels and radiator grilles, as well as in the electrical/electronic industry and telecommunications for televisions, personal computers, and telephones. This terpolymer is also used by producers of refrigeration equipment, toys, and sports equipment [1, 79]. ABS surfaces are ideal for further functionalization, e.g. by coating, galvanizing, hot embossing, or printing [79].

Figure 4.28 shows the pyrolysis–GC/MS chromatogram of ABS at 700°C. The degradation products identified by searching of the *NIST 05* mass spectra library are summarized in Table 4.10. The pyrolyzate of ABS contains all three monomers: 1,3-butadiene (t_R = 5.69 min), acrylonitrile (t_R = 5.84 min), and styrene (t_R = 9.58 min), which indicates that the

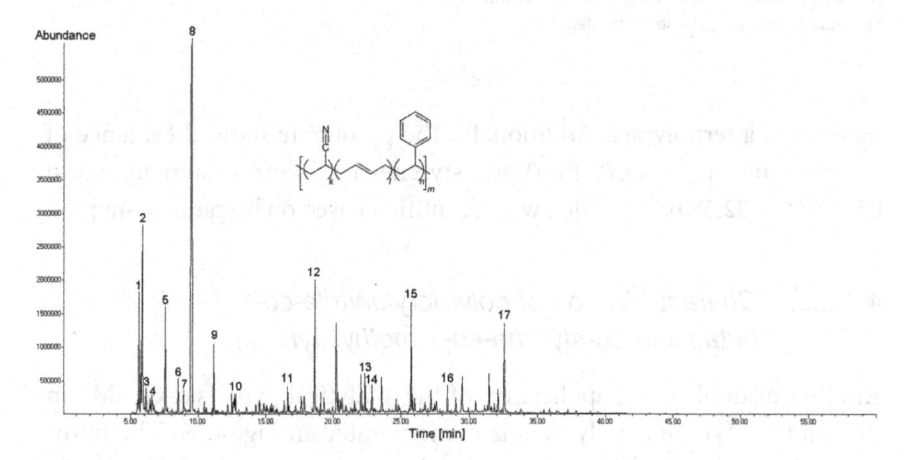

Fig. 4.28. Pyrolysis–GC/MS chromatogram of ABS at 700°C. Apparatus 1, GC column 1, GC conditions 2. For peak identification, see Table 4.10.

Table 4.10. Pyrolysis products of ABS at 700°C.

Peak No.	Retention time t_R (min)	Pyrolysis product
1	5.69	1,3-Butadiene
2	5.84	Acrylonitrile
3	6.12	Methacrylonitrile
4	6.55	Benzene
5	7.64	Toluene
6	8.52	4-Ethenylcyclohexene (1,3-Butadiene dimer)
7	8.92	Ethylbenzene
8	9.58	Styrene
9	11.16	(1-Methylethenyl)-benzene (α-Methylstyrene)
10	12.65	Indene
11	16.65	Benzenepropanenitrile
12	18.66	Benzenebutanenitrile
13	22.40	1,2-Diphenylethane
14	22.86	1,2-Diphenylpropane
15	25.70	Styrene dimer*
16	29.40	2,5-Diphenyl-1,5-hexadiene
17	32.58	2-Phenethyl-4-phenylpent-4-enenitrile (SAS, styrene–acrylonitrile-styrene hybrid trimer)*

Notes: Apparatus 1, GC column 1, GC conditions 2.
*Identification based on literature data [56].

material is a terpolymer. Additionally, the pyrolyzate showed evidence of styrene dimer (t_R = 25.70 min) and styrene–acrylonitrile–styrene hybrid trimer (t_R = 32.58 min), which were identified based on literature data [56].

4.5.2.3. *Characterization of poly(acrylonitrile-co-1,3-butadiene-co-styrene-co-α-methylstyrene)*

Another example of an application of the analytical pyrolysis–GC/MS in the field of styrene copolymers is demonstrated in Fig. 4.29. The pyrogram obtained shows the pyrolysis products of poly(acrylonitrile-*co*-1,3-butadiene-*co*-styrene-*co*-α-methylstyrene) (ABS-α-MS). The main pyrolytic

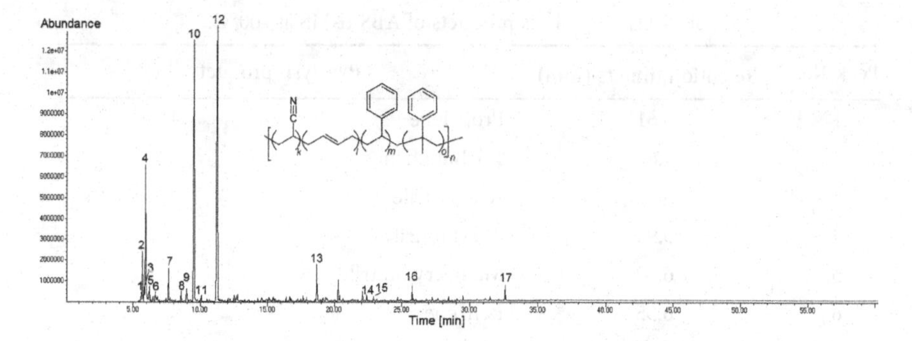

Fig. 4.29. Pyrolysis–GC/MS chromatogram of ABS-α-MS at 700°C. Apparatus 1, GC column 1, GC conditions 2. For peak identification, see Table 4.11.

degradation products of ABS-α-MS at 700°C are the monomers: 1,3-butadiene (t_R = 5.69 min), acrylonitrile (t_R = 5.92 min), styrene (t_R = 9.58 min), and α-methylstyrene (t_R = 11.16 min). The detailed identification of pyrolysis products of ABS-α-MS at 700°C is given in Table 4.11.

4.5.2.4. *Characterization of poly(styrene-co-divinylbenzene)*

Styrene–divinylbenzene copolymers (SDVB) are free flowing, spherical beads made from tough styrenic polymers that have been crosslinked with divinylbenzene [81]. They are sold as clear to colorless opaque spheres that are often used as plastic ball bearings. The copolymers are useful for many applications, such as lubrication, void space maintainers, grinding media, adsorption and decolorization, precision fillers, and lubrication [81]. Because of their hardness and spherical nature, they act like tiny, plastic ball bearings and have found use in many lubrication applications. One such application is wall drilling, where the copolymer is added to the drilling mud for enhanced lubrication in difficult rock formations and in horizontal drilling. SDVB copolymers are used as proppant in oil field fracturing due to their excellent crush strengths and natural void area. They are also useful as grinding media due to their good crush strengths and spherical nature. The copolymers have found application in sand blasting and deburring of metal and plastic parts [81].

Styrene–divinylbenzene copolymers are made of polymerized styrene, so they act as adsorption media for a wide variety of organic compounds. They can be particularly useful for removing higher MW oils. Because of

Table 4.11. Pyrolysis products of ABS-α-MS at 700°C.

Peak No.	Retention time t_R (min)	Pyrolysis product
1	5.61	Propylene
2	5.69	1,3-Butadiene
3	5.84	Acetonitrile
4	5.92	Acrylonitrile
5	6.20	Methacrylonitrile
6	6.55	Benzene
7	7.64	Toluene
8	8.52	4-Ethenylcyclohexene (Butadiene dimer)
9	8.92	Ethylbenzene
10	9.58	Styrene
11	10.08	Cumene
12	11.16	(1-Methylethenyl)-benzene (α-Methylstyrene)
13	18.66	Benzenebutanenitrile
14	22.40	1,2-Diphenylethane
15	22.90	α-Methylbibenzyl
16	25.70	Styrene dimer*
17	32.60	2-Phenethyl-4-phenylpent-4-enenitrile (SAS, styrene–acrylonitrile-styrene hybrid trimer)*

Notes: Apparatus 1, GC column 1, GC conditions 2.
*Identification based on literature data [56].

their relatively low cost, they are often used as chromatographic stationary phases in GC. Styrene–divinylbenzene copolymer is used as packing material (solid phase) to determine the contamination of groundwater by pesticides using solid phase extraction (SPE) method. It is widely used as column packing material in liquid chromatography for separation of both organic and inorganic ions. This copolymer is also converted into ion exchangers. SDVB copolymers are made from crosslinked PS, so they will not melt at elevated temperatures. The beads can be used as whole beads or ground and in a wide variety of filler applications. The ground copolymer is often used as a mold release agent.

Figure 4.30 shows the pyrolysis–GC/MS chromatogram of poly(styrene-*co*-divinylbenzene) (SDVB) at 700°C. The thermal degradation products of SDVB are summarized in Table 4.12.

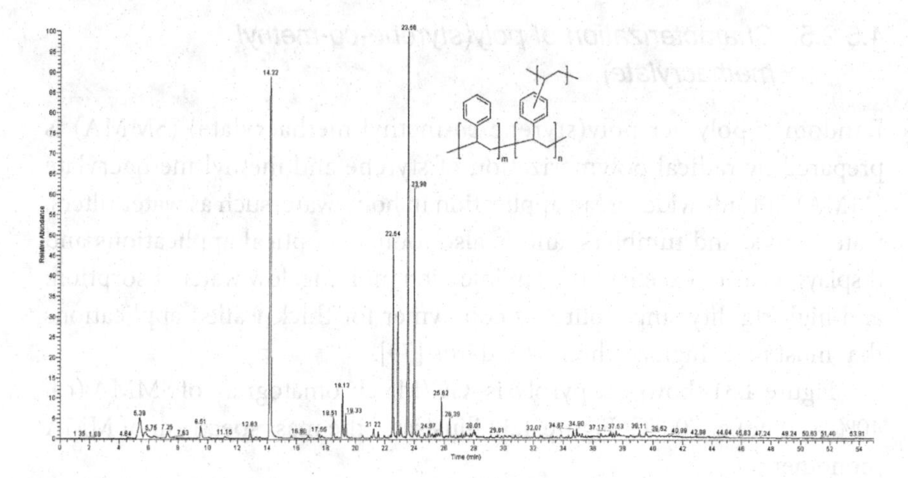

Fig. 4.30. Pyrolysis–GC/MS chromatogram of SDVB at 700°C. Apparatus 2, GC column 3, GC conditions 3. For peak identification, see Table 4.12.

Table 4.12. Pyrolysis products of SDVB at 700°C.

Retention time t_R (min)	Pyrolysis product
5.39	2-Methyl-1-propene
9.51	Toluene
12.83	Ethylbenzene
13.24	*p*-Xylene
14.22	Styrene
15.67	Cumene
16.80	2-Propenylbenzene
17.55	*o*-Ethyltoluene
18.51	*α*-Methylstyrene
19.12	*m*-Methylstyrene
19.33	*p*-Methylstyrene
21.22	Indene
22.54	*m*-Ethylstyrene
22.87	*p*-Ethylstyrene
23.56	*m*-Divinylbenzene
23.98	*p*-Divinylbenzene

Notes: Apparatus 2, GC column 3, GC conditions 3.

4.5.2.5. *Characterization of poly(styrene-co-methyl methacrylate)*

Random copolymer poly(styrene-*co*-methyl methacrylate) (SMMA) is prepared by radical polymerization of styrene and methyl methacrylate (MMA). It finds widespread application in homeware, such as water filters, water tanks, and tumblers, and is also found in optical applications and displays. Clarity, excellent flow in injection molding, low water absorption, and high rigidity single out this copolymer for thick-walled applications that must meet high aesthetic standards [79].

Figure 4.31 shows the pyrolysis–GC/MS chromatogram of SMMA (ca. 40% styrene) at 700°C. Figure 4.32 illustrates the mass spectrum of MMA monomer.

4.6. Analysis of Rubber Materials

Rubber is an elastic substance obtained from the exudations of certain tropical plants (natural rubber (NR)) or derived from petroleum and natural gas (synthetic rubber) [82]. Rubbers are frequently filled with opaque materials like carbon black, making them difficult to analyze

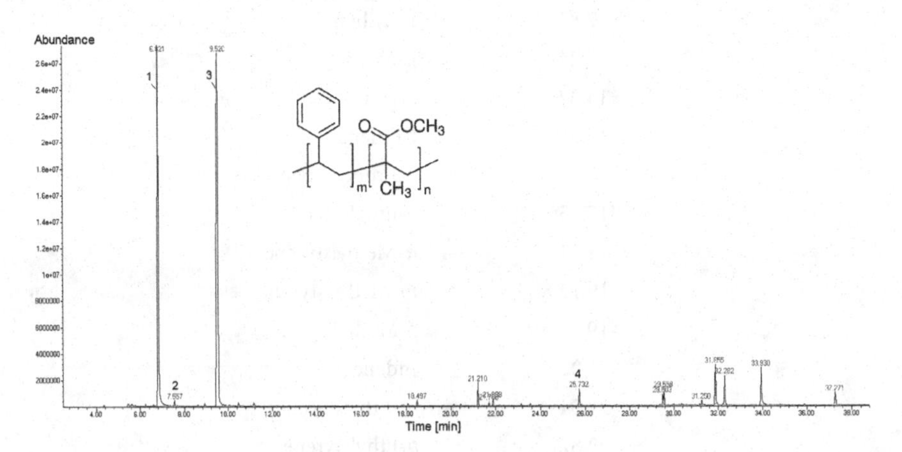

Fig. 4.31. Pyrolysis–GC/MS chromatogram of SMMA (ca. 40% styrene) at 700°C. Apparatus 1, GC column 1, GC conditions 2. Peak identification: (1) t_R = 6.82 min — MMA, (2) t_R = 7.56 — toluene, (3) t_R = 9.52 — styrene, (4) t_R = 25.73 min — styrene dimer.

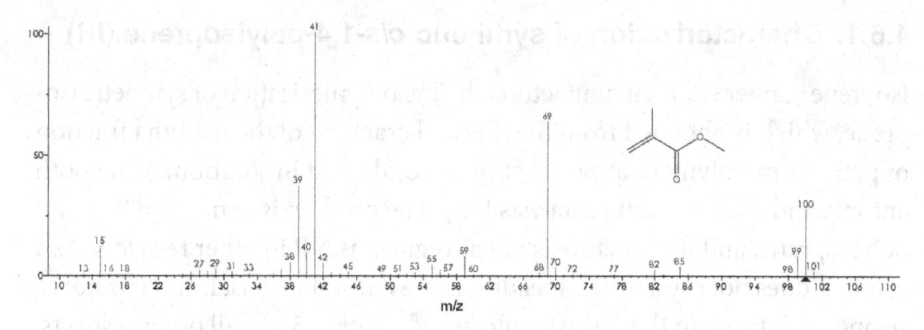

Fig. 4.32. Mass spectrum of MMA.

by spectroscopy. Furthermore, crosslinking makes them insoluble and thus many of the traditional analytical tools for organic analysis are difficult or impossible to apply. Therefore, the study of rubber is the oldest application of analytical pyrolysis [1]. The more commonly commercially used rubbers are NR (polyisoprene), synthetic polyisoprene (IR), polyisobutene (PIB), isobutylene–isoprene copolymer (butyl rubber, IIR), polybutadiene (butadiene rubber, BR), EPM, EPDM, poly(acrylonitrile-*co*-butadiene) (nitrile–butadiene rubber, NBR), polychloroprene (neoprene, CR), poly(styrene-*co*-butadiene) (styrene–butadiene rubber, SBR), PDMS (silicone rubber), and fluoroelastomere. Because of its elasticity, resilience, and toughness, rubber is the basic constituent of the tires used in automotive vehicles, aircrafts, and bicycles. More than half of all rubber produced goes into automobile tires; the rest goes into mechanical parts such as mountings, gaskets, belts, and hoses, as well as consumer products such as shoes, clothing, furniture, and toys [82]. Each rubber compound in a tire contains rubber polymers, sometimes one, but often a blend of two or more. SBR is widely used for tread compounds of a tire, generally in the tire treads of passenger cars, and NR in the relatively large-sized tires of buses and trucks etc. Inner tubes are usually based on butyl rubber, a copolymer of isobutylene with a small proportion of isoprene (IR) [83–84]. Pyrolysis of some kinds of rubber like EPM, EPDM, and fluoroelastomers are already described in Sections 4.3.1, 4.3.6, and 4.4.4, respectively. The other kinds of rubber are characterized in the following text.

4.6.1. Characterization of synthetic *cis*-1,4-polyisoprene (IR)

Isoprene rubber (IR) is manufactured by the polymerization of synthetic isoprene, which is obtained from the thermal cracking of the naphtha fraction of petroleum Polymerization reaction is conducted in solution, using both anionic and Ziegler–Natta catalysts [85]. The product is at most 98% *cis*-1,4 polyisoprene, and its structure is not as regular as NR in other respects. As a result, it does not crystallize as readily as the natural material, and it is not as strong or as tacky in the raw (unvulcanized) state [85]. In all other respects, though, IR is a complete substitute for NR. Currently synthetic polyisoprene is being used in a wide variety of industries, in applications requiring low water swell, high gum tensile strength, good resilience, high hot tensile and good tack [86]. Gum compounds based on synthetic polyisoprene are being used in rubber bands, cut thread, baby bottle nipples, and extruded hose. Black-loaded compounds find use in tires, motor mounts, pipe gaskets, shock absorber bushings and many other molded and mechanical goods. Mineral-filled systems find applications in footwear, sponge, and sporting goods [86].

Figure 4.33 shows the pyrolysis–GC/MS chromatogram of synthetic *cis*-1,4-polyisoprene (IR) at 700°C. The pyrolysis products of synthetic polyisoprene, identified by using the *NIST 05* mass spectra library are summarized in Table 4.13. As can be seen from Fig. 4.33 and from Table 4.13, the main volatile products formed from pyrolysis of *cis*-1,4- polyisoprene are isoprene and isoprene dimer species (dipentene). Dipentene comprises of the racemic mixture of the two enantiomers D- and L-limonene. Its systematic name is 1-methyl-4-(1-methylethenyl)-cyclohexene, an unsaturated hydrocarbon classified as a monoterpene. It is the dimer of two isoprene units, which originate from the (either natural or synthetic) polyisoprene. Figure 4.34 shows the chemical structures of polyisoprene, dipentene, and isoprene [87].

4.6.2. Characterization of polyisobutene (polyisobutylene, PIB)

PIB is a synthetic rubber or elastomer. PIB, sometimes called *butyl rubber*, is made from the monomer isobutylene (2-methyl-1-propene) by cationic polymerization [88]. PIB was first developed during the early 1940s. At that

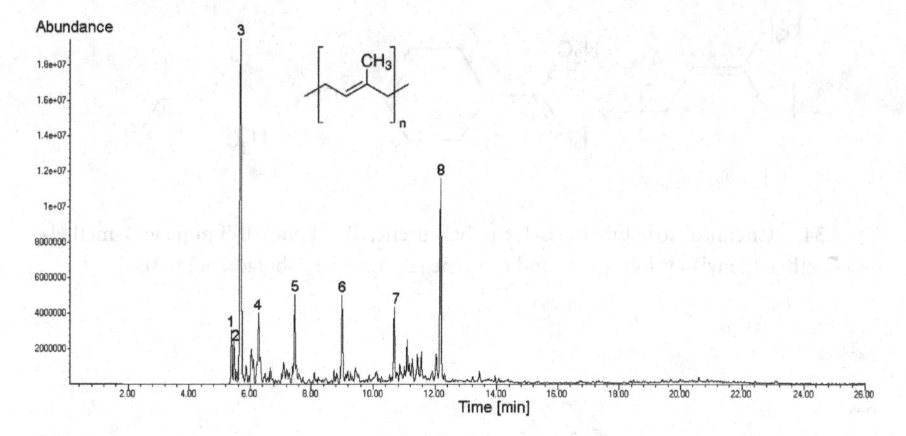

Fig. 4.33. Pyrolysis–GC/MS chromatogram of synthetic *cis*-1,4-polyisoprene (IR) at 700°C. Apparatus 1, GC column 1, GC conditions 2. For peak identification, see Table 4.13.

Table 4.13. Pyrolysis products of synthetic *cis*-1,4-polyisoprene (IR) at 700°C.

Peak No.	Retention time t_R (min)	Pyrolysis product
1	5.39	Propylene
2	5.47	2-Butene
3	5.68	2-Methyl-1,3-butadiene (Isoprene)
4	6.27	5-Methyl-1,3-cyclopentadiene
5	7.47	Toluene
6	9.00	*p*-Xylene
7	10.69	1,5-Dimethyl-1,5-cyclooctadiene
8	12.18	dl-Limonene (Dipentene, isoprene dimer)

Notes: Apparatus 1, GC column 1, GC conditions 2.

time, the most widely used rubber was NR (polyisoprene). PIB is the only rubber that is gas impermeable and can hold air for long periods of time. Because PIB will hold air, it is used to make things like the inner liner of tires and the inner liners of basketballs [88].

Figure 4.35 shows the pyrolysis–GC/MS chromatogram of polyisobutene (PIB) at 700°C. The identified pyrolysis products of PIB are summarized in Table 4.14.

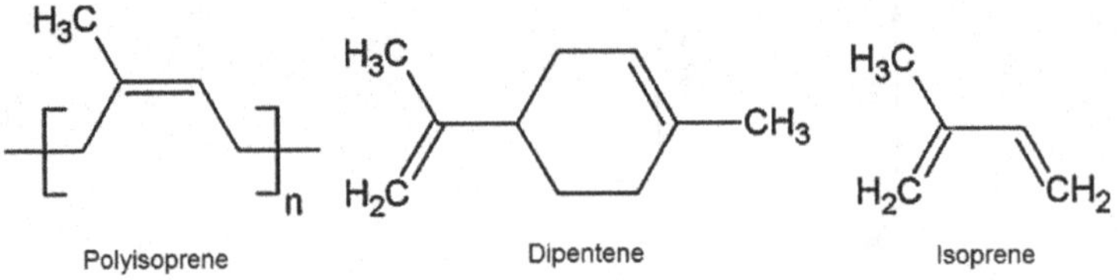

Polyisoprene Dipentene Isoprene

Fig. 4.34. Chemical structure of *cis*-1,4-polyisoprene, dipentene (dl-limonene, 1-methyl-4-(1-methylethenyl)-cyclohexene), and isoprene (2-methyl-1,3-butadiene) [87].

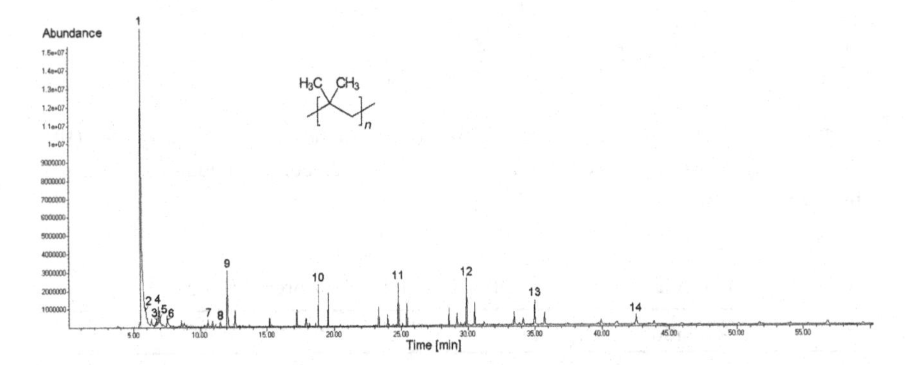

Fig. 4.35. Pyrolysis–GC/MS chromatogram of polyisobutene (PIB) at 700°C. Apparatus 1, GC column 1, GC conditions 2. For peak identification, see Table 4.14.

4.6.3. Characterization of poly(isobutene-*co*-isoprene) (butyl rubber, IIR)

Butyl rubber is a copolymer of isobutylene with a small amount of isoprene (1~2 wt.%) and has a wide range of applications including inner tire tubes, cable insulation, vibration dampers, pharmaceutical stoppers, automotive parts, protective clothing, and gas masks [89]. The pyrogram of IIR is similar to the pyrogram of polybutylene (PIB), showed in Fig. 4.35. The small amount of isoprene is difficult to detect.

4.6.4. Characterization of polybutadiene (butadiene rubber, BR)

Polybutadiene is a homopolymer (only one monomer) of 1,3-butadiene. For a typical BR, MW is usually >100,000 grams per mole. This represents

Table 4.14. Pyrolysis products of polyisobutene (PIB) at 700°C.

Peak No.	Retention time t_R (min)	Molar mass (g/mol)	Pyrolysis product
1	5.55	56	2-Methyl-1-propene
2	6.02	56	2-Butene
3	6.72	98	2,4-Dimethyl-1,3-pentadiene
4	6.85	112	2,4,4-Trimethyl-1-pentene
5	6.95	112	2,4,4-Trimethyl-2-pentene
6	7.52	128	2,2,4,4-Tetramethylpentane
7	10.58	154	2,4,4,6-Tetramethyl-2-heptene
8	11.52	168	2,2,4,6,6-Pentamethyl-3-heptene
9	12.06	168	2,4,4,6,6-Pentamethyl-2-heptene*
10	18.79	224	2,4,4,6,6,8,8-Heptamethyl-2-nonene*
11	24.76	280	2,4,4,6,6,8,8,10,10-Nonamethyl-2-undecene*
12	29.89	336	2,4,4,6,6,8,8,10,10,12,12-Undecamethyl-2-tridecene*
13	34.98	392	2,4,4,6,6,8,8,10,10,12,12,14,14-Tridecamethyl-2-pentadecene*
14	42.54	448	2,4,4,6,6,8,8,10,10,12,12,14,14,16,16-Pentadecamethyl-2-heptadecene*

Notes: Apparatus 1, GC column 1, GC conditions 2.
*Identification based on literature data [30].

a chain that contains over 2,000 butadiene units [90]. Most BRs are made by a solution process, using either a transition metal (Nd, Ni, or Co) complex or an alkyl metal, like butyllithium, as catalyst. Since the reaction is very exothermic and can be explosive, particularly with alkyllithium catalysts, the reaction is normally carried out in solvents like hexane, cyclohexane, benzene, or toluene. The solvents are used to reduce the rate of reaction, control the heat generated by the polymerization, and to lower the viscosity of the polymer solution in the reactor [90].

Polybutadiene (BR) is the second largest volume synthetic rubber produced, next to SBR. The major use of BR is in tires with over 70% of the polymer produced going into treads and sidewalls. Cured BR imparts excellent abrasion resistance (good tread wear) and low rolling resistance (good fuel economy) due to its low glass transition temperature. BR is usually blended with other elastomers like NR or SBR for tread compounds [90]. BR also has a major application as an impact modifier for PS and acrylonitrile–butadiene–styrene resin (ABS) with about 25% of the total volume going into these applications. About 20,000 metric tons of polybutadiene worldwide is used each year in golf ball cores due to its outstanding resiliency. Polybutadiene is and will continue to be a high-volume rubber for the use in tires, toughened plastics, and golf balls due to its low cost, availability, and unique properties [90].

Figure 4.36 presents the pyrolysis–GC/MS chromatogram of polybutadiene (BR) with additives at 700°C. The identified thermal degradation products of BR with additives are summarized in Table 4.15.

4.6.5. Characterization of poly(acrylonitrile-*co*-1,3-butadiene) (nitrile rubber, NBR)

Poly(acrylonitrile-*co*-1,3-butadiene) (NBR) is a copolymer containing 15–50% acrylonitrile, manufactured by emulsion polymerization of acrylonitrile and 1,3-butadiene. It was invented around the same time as styrene–butadiene copolymers in the German program (at the end of the 1920s) to find substitutes for NR [91]. The major applications for this material are in areas requiring oil and solvent resistance. The largest market for nitrile rubber is in the automotive industry because of its solvent and oil resistance. Major end uses are for hoses, fuel lines, O-rings, gaskets, and

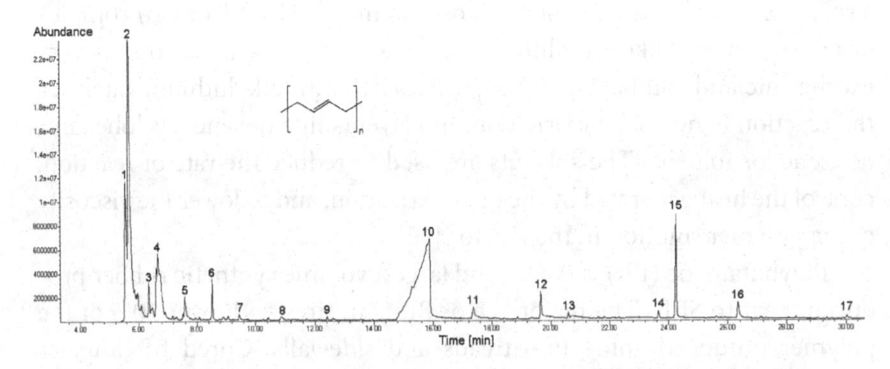

Fig. 4.36. Pyrolysis–GC/MS chromatogram of polybutadiene (BR) with additives at 700°C. Apparatus 1, GC column 1, GC conditions 2. For peak identification, see Table 4.15.

Table 4.15. Pyrolysis products of polybutadiene (BR) with additives at 700°C.

Peak No.	Retention time t_R (min)	Pyrolysis product	Pyrolyzed material
1	5.55	Propylene	BR
2	5.66	1,3-Butadiene	BR
3	6.35	Tetrahydrofuran	Solvent/Additive
4	6.66	Benzene	BR
5	7.58	Toluene	BR
6	8.52	4-Ethenylcyclohexene (1,3-Butadiene dimer)	BR
7	9.45	Styrene	BR
8	10.88	Benzaldehyde	Additive
9	12.41	3-Butenylbenzene	BR
10	15.87	Benzoic acid	Additive
11	17.37	4-Methylbenzoic acid	Additive
12	19.66	Biphenyl	BR
13	20.60	Diphenylmethane	BR
14	23.65	Fluorene	BR
15	24.25	Benzophenone	Additive
16	26.34	Fluorenone	Additive
17	30.03	Triphenylmethane	BR

Notes: Apparatus 1, GC column 1, GC conditions 2.

seals. In blends with PVC and poly(acrylonitrile-*co*-butadiene-*co*-styrene), nitrile rubber acts as an impact modifier. Some nitrile rubber is sold in latex form for the production of grease-resistant tapes, gasketing material, and abrasive papers. Latex also is used to produce solvent-resistant gloves [91]. Nitrile rubber is often used as the seal material for automotive and petroleum applications on account of its good resistance toward oil and solvent, and low gas permeability. Its aging resistance is of great importance, and thus concerned because of the unsaturated backbone of the butadiene part [92].

Figure 4.37 shows the pyrolysis–GC/MS chromatogram of NBR at 700°C. The pyrolysis products of this copolymer, identified by using the *NIST 05* mass spectra library, are summarized in Table 4.16. As can be seen from Fig. 4.37 and from Table 4.16, the main volatile products formed

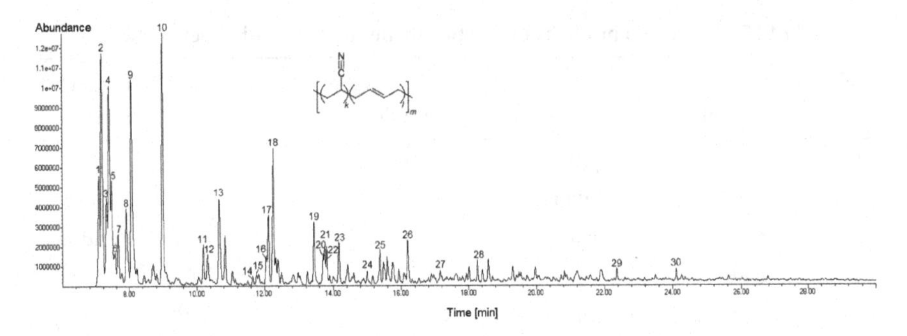

Fig. 4.37. Pyrolysis–GC/MS chromatogram of poly(acrylonitrile-*co*-1,3-butadiene) (NBR) at 700°C. Apparatus 1, GC column 1, GC conditions 1. For peak identification, see Table 4.16.

from pyrolysis of poly(acrylonitrile-*co*-1,3-butadiene) are the monomers 1,3-butadiene (Fig. 4.37, t_R = 7.16 min) and acrylonitrile (Fig. 4.37, t_R = 7.39 min). The presence of benzonitrile (Fig. 4.37, t_R = 12.26 min) and *p*-tolunitrile (Fig. 4.37, t_R = 14.17 min) in pyrogram of NBR is also characteristic for the pyrolysis of this copolymer. The detected triallyl isocyanurate (TAIC) (Fig. 4.37, t_R = 24.10 min) is a crosslinking agent to improve crosslinking efficiency by manufacturing of polymers and rubbers. The rubber obtained by using TAIC has better mechanical characteristics and larger heat, hydrolytic, and weather resistance.

4.6.5.1. *Identification of organic additives in nitrile rubber materials*

The criteria for assessing the quality of rubber materials are the polymer or copolymer composition and the additives. Commercial plastics and rubbers always contain low-MW additives. These compounds are important in ensuring the end-use properties of a polymer or copolymer [47]. Additives can improve or modify the mechanical properties (fillers and reinforcements), modify the color and appearance (pigments and dyestuffs), give resistance to heat degradation (antioxidants and stabilizers), provide resistance to light degradation (UV stabilizers), improve the flame resistance (flame retardants), improve the performance (antistatic or conductive additives, plasticizers, blowing agents, lubricants, mold

Table 4.16. Pyrolysis products of poly(acrylonitrile-*co*-1,3-butadiene) (NBR) at 700°C.

Peak No.	Retention time t_R (min)	Pyrolysis product
1	7.08	Propylene
2	7.16	1,3-Butadiene/2-Butene
3	7.31	Acetonitrile
4	7.39	Acrylonitrile
5	7.46	1,3-Cyclopentadiene
6	7.56	Propanenitrile
7	7.66	2-Methyl-2-propenenitrile
8	7.92	1,3-Cyclohexadiene
9	8.07	Benzene
10	8.99	Toluene
11	10.19	Ethylbenzene
12	10.32	*p*-Xylene
13	10.66	Styrene
14	11.66	Propylbenzene
15	11.77	1-Ethyl-3-methylbenzene
16	12.06	Aniline
17	12.12	*α*-Methylstyrene
18	12.26	Benzonitrile
19	13.43	Indene
20	13.70	Acetophenone
21	13.75	2-Methylbenzonitrile (*o*-Tolunitrile)
22	13.80	*p*-Toluidine
23	14.17	4-Methylbenzonitrile (*p*-Tolunitrile)
24	15.00	Benzylnitrile
25	15.38	1,2-Dihydronaphthalene
26	16.19	Naphthalene
27	17.16	2-Phenyl-2-propenenitrile
28	18.26	2-Methylnaphthalene
29	22.37	1-Isocyanonaphthalene
30	24.10	TAIC

Notes: Apparatus 1, GC column 1, GC conditions 1.

release agents, surfactants, and preservatives), and improve the processing characteristics (recycling additives) of polymers or copolymers [45–47]. Some of the additives accumulate in the environment and affect our health and the environment. Knowledge of additives is important for evaluating the environmental impact and interaction of polymeric materials, investigating long-term properties and degradation mechanisms, verifying ingredients, investigating manufacturing problems, qualifying control polymeric materials, identifying odorants, avoiding workplace exposure, and insuring safety of food packaging and medical products [93]. Identification of polymer or copolymer additives is also desired if competitor products are investigated. The quantification of additives is important for quality control and troubleshooting of the manufacturing processes. Both identification and quantification are difficult tasks because there is a wide variety of different additives, usually mixtures of additives are used, and the added amount is often low and can be further decreased because of degradation [94]. Most analytical methods reported for the determination of polymer or copolymer additives require previous extraction of the additives from the polymeric material. For this purpose, widely used sample preparation techniques such as liquid extraction with various solvents, Soxhlet extraction, ultrasonic-assisted extraction and pressurized liquid extraction, like microwave assisted extraction, accelerated solvent extraction, supercritical fluid extraction, and extraction processes using autoclaves are used [94]. All of these methods are very laborious and time-consuming. Subsequent analysis of the extracted additives has been performed using thin-layer chromatography (TLC), supercritical fluid chromatography (SFC), GC, high performance liquid chromatography (HPLC), and capillary electrophoresis (CE), or their conjunction with MS (GC/MS, HPLC–MS, and CE–MS) [94]. Infrared- or UV-spectroscopic investigations of polymers or copolymers, particularly in the form of a thin film, are only applicable for a sample containing just a single additive or if determination of a sum parameter is sufficient. Several reports on the use of matrix-assisted laser desorption–ionization mass spectrometry (MALDI-MS) for the detection of additives in plastic samples exist [94]. Nevertheless, these methods require fine grinding of the polymer or extraction and dissolution of the sample and subsequent analysis of the extract. Besides investigations on thin polymer films using time-of-flight secondary ion mass spectrometry

(TOF-SIMS), polymer additives have been detected directly in polymer samples by laser-MS. The direct MS methods are the focus of current research for identification of solid polymeric materials [94]. The thermo-analytical techniques, such as TGA or temperature-programmed analytical pyrolysis, specifically take advantage of relatively slow heating, particularly in combination with appropriate detection modes like thermogravimetric-MS, thermogravimetric-FTIR spectroscopy, and temperature-programmed pyrolysis–GC/MS [47]. Headspace solid-phase microextraction and thermal extraction in conjunction with GC/MS have been also applied for the extraction and identification of several common polymer or copolymer additives [47]. In such volatile removal techniques, the additives are usually detected at temperatures below the decomposition temperature of the polymer or copolymer. It is also possible to gain information about additives from the EGA–MS technique (Evolved gas analysis-MS) [16–19, 56]. In EGA–MS, evolved gases formed during the programmed heating of the sample are directly transferred into MS to achieve online monitoring of the components. For the EGA–MS measurements, the separation column in Py–GC/MS is replaced with a deactivated open transfer line, connecting directly between a temperature programmable pyrolyzer and an ion source of MS. The resulting thermogram monitored by MS during the programmed heating reflects the evolved gas profile of the sample as a function of temperature. The observed specific pyrograms and/or thermograms often provide valuable information regarding the composition and/or chemical structures of the original polymer sample as well as the degradation mechanisms and related kinetics [56].

The flash analytical pyrolysis technique hyphenated to GC/MS has extended the range of possible tools for characterization of synthetic polymers or copolymers. In previous works of the author (P. K.), nitrile rubber materials were studied using direct analytical flash pyrolysis hyphenated to GC and electron impact ionization MS, in both scan and SIM modes to demonstrate that this technique is a good tool to identify the organic additives in nitrile rubber [12, 47, 95, 96]. Figure 4.38 shows the mass spectra of additives or their thermal decomposition products, identified in two investigated nitrile rubber (NBR) materials. The pyrograms of the NBR materials are shown in earlier publication [47] of the author (P. K.). The identification of pyrolyzates of poly(acrylonitrile-*co*-1,3-butadiene) (NBR) samples

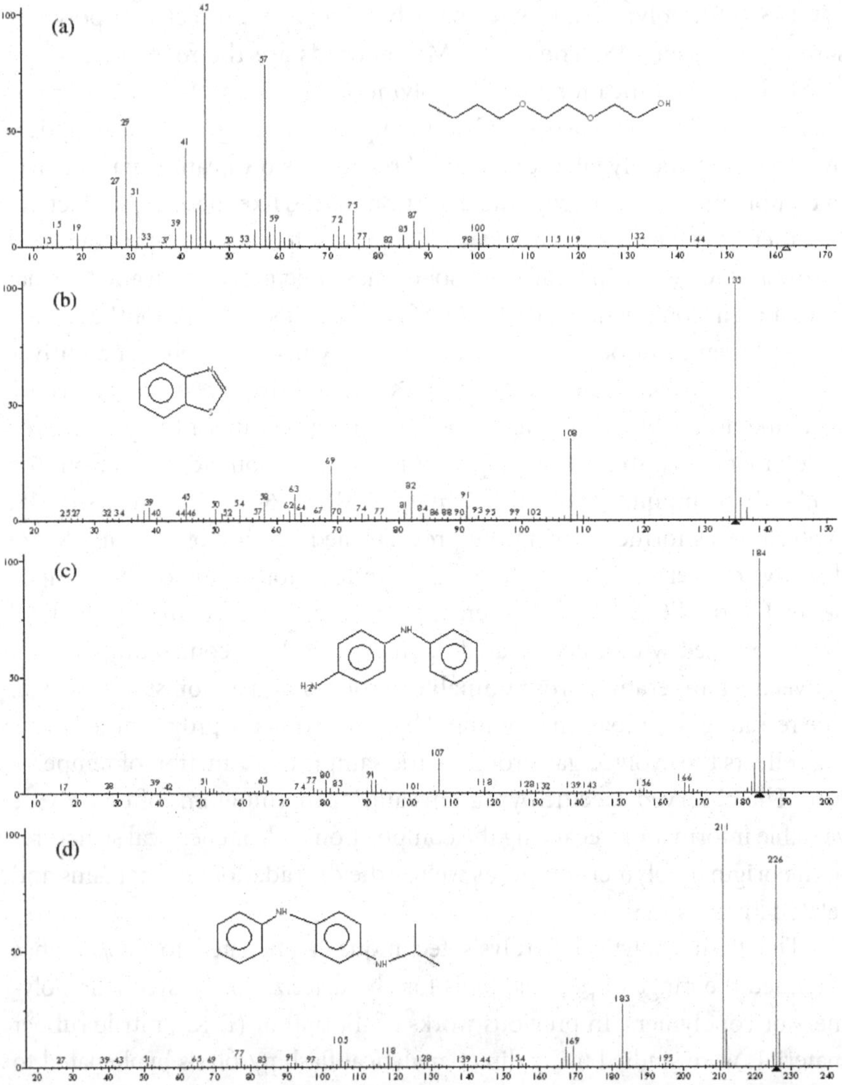

Fig. 4.38. Mass spectra of additives identified in two investigated nitrile rubber (NBR) materials: (a) diethylene glycol mono-*n*-butyl ether, (b) benzothiazole, (c) *N*-phenyl-1,4-benzenediamine, (d) *N*-(1-methylethyl)-*N*′-phenyl-1,4-benzenediamine.

Table 4.17. Pyrolysis products of the investigated nitrile rubber (NBR) materials at 700°C.

Retention time t_R (min)	Pyrolysis product
6.67	Propylene
6.76	1,3-Butadiene
6.98	Acrylonitrile
7.05	1,3-Cyclopentadiene
7.24	Methacrylonitrile
7.50	1,4-Cyclohexadiene
7.66	Benzene
8.57	Toluene
8.73	2-Ethyl-1-hexene
10.25	Styrene
11.83	Benzonitrile
12.38	2-Ethyl-1-hexanol
15.36	Diethylene glycol mono-*n*-butyl ether
16.61	Benzothiazole
18.24	Phthalic anhydride
25.52	2-Ethylhexyl benzoate
29.96	*N*-phenyl-1,4-benzenediamine
32.35	*N*-(1-methylethyl)-*N'*-phenyl-1,4-benzenediamine

Notes: Apparatus 1, GC column 2, GC conditions 1. For pyrolysis–GC/MS chromatograms (pyrograms), see Ref. [47].

indicated in Table 4.17 shows the presence of compounds generated at 700°C from the acrylonitrile sequences (acrylonitrile, methacrylonitrile, benzonitrile), from the butadiene sequences (1,3-butadiene), and from the additives of the copolymers, like 2-ethyl-1-hexene (t_R = 8.73 min), 2-ethyl-1-hexanol (t_R = 12.38 min), phthalic anhydride (t_R = 18.24 min), 2-ethylhexyl benzoate (t_R = 25.52 min), diethylene glycol mono-*n*-butyl ether (t_R = 15.36 min), benzothiazole (t_R = 16.61 min), *N*-phenyl-1,4-benzenediamine (t_R = 29.96 min), and *N*-(1-methylethyl)-*N'*-phenyl-1,4-benzenediamine (t_R = 32.35 min). The identified substance diethylene glycol mono-*n*-butyl ether (2-[2-butoxyethoxy]-ethanol, CAS No. 112-34-5) (t_R = 15.36 min), is a residual solvent from the rubber sample. The substance is known as an excellent coalescing and coupling agent. The identified benzothiazole (t_R = 16.61 min) has been formed by the thermal degradation of 2-mercaptobenzothiazole (CAS No. 149-30-4). 2-Mercaptobenzothiazole is used as an accelerator for the vulcanization of rubber and as an antioxidant. The identified *N*-(1-methylethyl)-*N*′-phenyl-1,4-benzenediamine (*N*-isopropyl-*N*′-phenyl-*p*-phenylendiamine, CAS number 101-72-4) (t_R = 32.35 min) is a very effective antioxidant and antiozonant that provides medium- to long-term protection for all synthetic and NR. Furthermore, *N*-phenyl-1,4-benzenediamine (t_R = 29.96 min) was probably generated from *N*-(1-methylethyl)-*N*′-phenyl-1,4-benzenediamine during the pyrolysis of the nitrile rubber sample. Another additive identified in nitrile rubbers was the plasticizer BEHP (CAS number 117-81-7). Phthalate esters, also known as phthalates or phthalic acid esters (PAEs), are a large group of synthetic organic compounds. They were introduced commercially in the 1920s and have been used as plasticizers in industrial and household products for over 50 years [97]. PAEs can be chemically classified into two groups: (1) low-molecular-weight (LMW) PAEs, which are mainly used as solvents to stabilize the fragrance and color in detergents, perfumes, and personal care products, and (2) high-molecular-weight (HMW) PAEs that are widely used in PVC, rubber and food packaging to increase the plasticity, flexibility, and elasticity [97]. The thermal decomposition of the plasticizer BEHP leads to the formation of 2-ethyl-1-hexene (t_R = 8.73 min), 2-ethyl-1-hexanol (t_R = 12.38 min), phthalic anhydride (t_R = 18.24 min), and 2-ethylhexyl benzoate (t_R = 25.52 min) during the pyrolysis [12, 47, 73, 74]. The EI mass spectra of the compounds and the appropriate chemical structures are shown in Fig. 4.20 in Section 4.4.1.

4.6.6. Characterization of poly(styrene-*co*-1,3-butadiene) (SBR)

SBR is derived from two monomers, styrene and 1,3-butadiene. The mixture of these two monomers is polymerized by two processes — from solution (S-SBR) or as an emulsion (E-SBR). E-SBR is more widely used [82]. E-SBR produced by emulsion polymerization is initiated by free radicals. S-SBR is produced by an anionic polymerization process. About 50% of car tires are made from various types of SBR. The styrene/butadiene ratio influences the properties of the polymer. With high styrene content, the rubbers are harder and less rubbery. SBR is not to be confused with a thermoplastic elastomer made from the same monomers, styrene–butadiene block copolymer. Styrene–butadiene copolymers are still almost exclusively blended with commodity PS, primarily intended for transparent, rigid packaging for foodstuffs, and consumer goods [79]. Food packaging continues to be an application for styrene–butadiene copolymers in multi-layer composite films with other polymers, some of which provide functional tasks, such as gas barrier properties.

Figure 4.39 illustrates the pyrolysis–GC/MS chromatogram of poly(styrene-*co*-1,3-butadiene) (SBR) at 700°C. The identified pyrolysis products of SBR are summarized in Table 4.18.

4.6.7. Characterization of polychloroprene (neoprene, CR)

Polychloroprene [poly(2-chloro-1,3-butadiene)] (CR) is one of the most important elastomers with an annual consumption of nearly 300,000 tons worldwide [98]. First production was in 1932 by DuPont (*Duprene*®, later *Neoprene*®), and since then CR has an outstanding position due to its technical properties. CR has good mechanical strength, high ozone and weather resistance, good aging resistance, low flammability, good resistance toward chemicals, moderate oil and fuel resistance, and adhesion to many substrates [98]. From the beginning until the 1960s, chloroprene was produced by

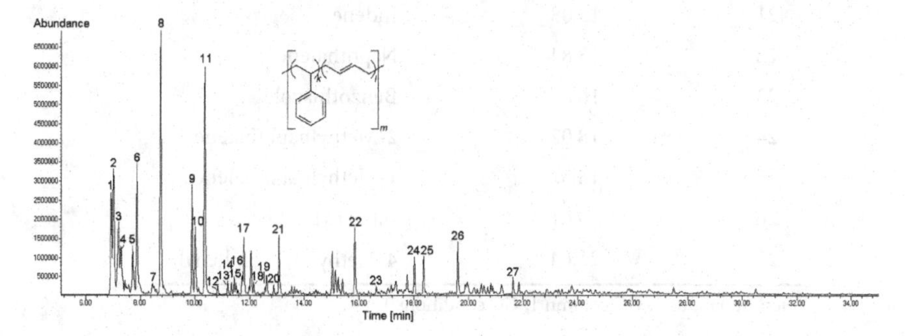

Fig. 4.39. Pyrolysis–GC/MS chromatogram of poly(styrene-*co*-1,3-butadiene) (SBR) at 700°C. Apparatus 1, GC column 2, GC conditions 1. For peak identification, see Table 4.18.

Table 4.18. Pyrolysis products of poly(styrene-*co*-1,3-butadiene) (SBR) at 700°C.

Peak No.	Retention time t_R (min)	Pyrolysis product
1	6.94	Propylene
2	7.02	1,3-Butadiene/2-Butene
3	7.21	1,3-Pentadiene
4	7.30	1,3-Cyclopentadiene
5	7.73	1,4-Cyclohexadiene
6	7.88	Benzene
7	8.45	1-Methyl-1,4-cyclohexadiene
8	8.75	Toluene
9	9.90	Ethylbenzene
10	10.02	*p*-Xylene
11	10.37	Styrene
12	10.84	Isopropylbenzene (Cumene)
13	11.20	2-Propenylbenzene
14	11.34	Propylbenzene (Isocumene)
15	11.44	1-Ethyl-3-methylbenzene
16	11.50	1-Ethyl-4-methylbenzene
17	11.79	α-Methylstyrene
18	12.14	*o*-Methylstyrene
19	12.57	1,2,3-Trimethylbenzene
20	12.89	Indane
21	13.08	Indene
22	15.84	Naphthalene
23	16.62	Benzothiazole
24	18.02	2-Methylnaphthalene
25	18.37	1-Methylnaphthalene
26	19.61	Biphenyl
27	21.64	4-Methyl-1,1′-biphenyl

Notes: Apparatus 1, GC column 1, GC conditions 1.

the older "acetylene process." This process has the disadvantages of being very energy intensive and having high investment costs. The modern chloroprene process, which is now used by nearly all producers, is based on butadiene, which is readily available. Butadiene is converted into the monomer 2-chloro-1,3-butadiene (chloroprene) via 3,4-dichloro-1-butene. In principle it is possible to polymerize chloroprene by anionic, cationic, and Ziegler–Natta catalysis techniques. However, because of product properties and economic considerations, free radical emulsion polymerization is used exclusively today. It is carried out in a commercial scale using both batch and continuous processes [98]. Polychloroprene can be vulcanized by using various accelerator systems over a wide temperature range. CR is used in different technical areas, mainly in the rubber industry (ca. 61%), but is also important as a raw material for adhesives (both solvent based and water based, ca. 33%) and has different latex applications (ca. 6%) such as dipped articles (e.g. gloves), moulded foam, and improvement of bitumen. Application areas in the elastomer field are widely spread, such as moulded goods, transmission belts, conveyor belts, profiles, gaskets, cable jackets, tubing, seals, O-rings, tire-sidewalls, gasoline hoses, and weather-resistant products such as wet suits and orthopedic braces. It is also used as a base resin in adhesives, electrical insulations, and coatings [98, 99].

Figure 4.40 shows the pyrolysis–GC/MS chromatogram of polychloroprene (CR) at 700°C. The pyrolysis products of polychloroprene at 700°C are shown in Table 4.19.

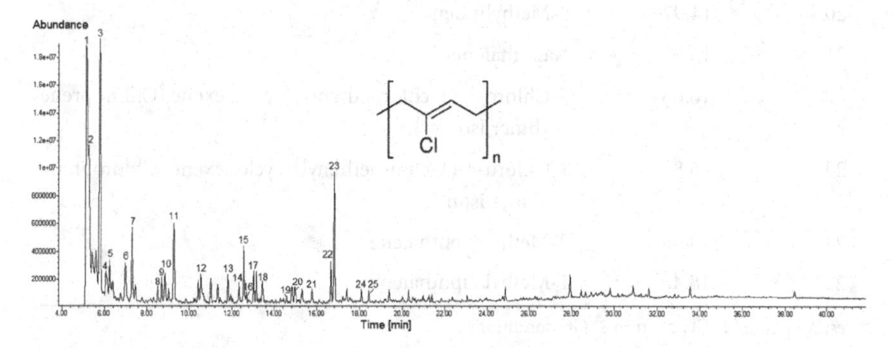

Fig. 4.40. Pyrolysis–GC/MS chromatogram of polychloroprene (CR) at 700°C. Apparatus 1, GC column 1, GC conditions 2. For peak identification, see Table 4.19.

Table 4.19. Pyrolysis products of polychloroprene (CR) at 700°C.

Peak No.	Retention time t_R (min)	Pyrolysis product
1	5.15	Hydrogen chloride (HCl)
2	5.26	1-Butene
3	5.78	2-Chloro-1,3-butadiene (Chloroprene monomer)
4	6.08	1,4-Cyclohexadiene
5	6.25	Benzene
6	6.99	1,1-Dichloro-1,2-butadiene
7	7.30	Toluene
8	8.53	Chlorobenzene
9	8.71	Ethylbenzene
10	8.87	*p*-Xylene
11	9.29	Styrene
12	10.55	1-Chloro-4-methylbenzene
13	11.81	1-Methyl-2-(1-methylethyl)benzene
14	12.33	1-Ethenyl-2-methylbenzene
15	12.56	Indene
16	12.68	1-Chloro-4-ethylbenzene
17	13.02	1-Chloro-4-ethenylbenzene
18	13.44	4-Chloro-1,2-dimethylbenzene
19	14.60	4-Ethenyl-1,2-dimethylbenzene
20	14.97	3-Methylindene
21	15.77	Naphthalene
22	16.66	1-Chloro-5-(1-chloroethenyl)-cyclohexene (Chloroprene dimer isomer)
23	16.82	1-Chloro-4-(1-chloroethenyl)-cyclohexene (Chloroprene dimer isomer)
24	18.08	2-Methylnaphthalene
25	18.43	1-Methylnaphthalene

Notes: Apparatus 1, GC column 1, GC conditions 2.

4.6.8. Characterization of PDMS (silicone rubber)

The silicones are known as polyorganosiloxanes according to IUPAC rules. Linear and cyclic polyorganosiloxanes are generally produced by reacting organodichlorosilanes with water [1, 96]. The most important industrial polyorganosiloxanes are the linear polydimethylsiloxanes (PDMSs) of the structure $-[(CH_3)_2SiO]_n-$. The polymers are classified according to their viscosity and the nature of the end-groups. The end-groups determine the use. For example, trimethylsilyl-terminated PDMSs are typical silicone fluids. Hydroxy- and vinyl-terminated polymers find major application in silicone rubbers [1, 96]. Silicone fluids are used as heat-transfer media in the chemical, petrochemical, pharmaceutical, and food industries. They are also used in solar power and climate simulation plants. Silicone fluids, especially functional copolymers with Si–H and Si–OH groups, have many uses as waterproofing agents for textiles and as protective coating for building materials. Silicone elastomers are used in the electrical, electronic, and automobile industries as cable sheaths, coatings, cable guides, and connectors and airbags as well as in medicine and dentistry as membranes, lenses, catheters, and implants [1, 96]. Figure 4.41 shows the Py–GC/MS chromatogram

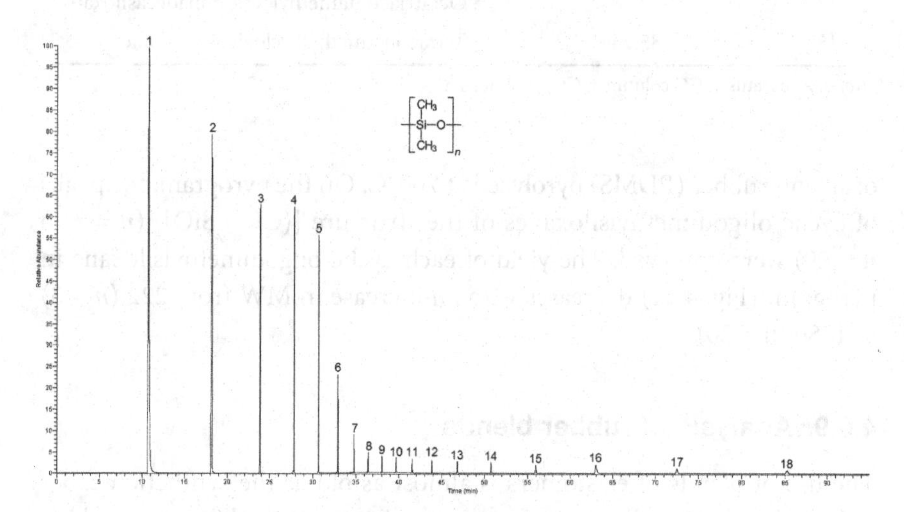

Fig. 4.41. Pyrolysis–GC/MS chromatogram of PDMS at 700°C. Apparatus 2, GC column 3, GC conditions 3. For peak identification, see Table 4.20.

Table 4.20. Pyrolysis products of PDMS at 700°C.

Peak No.	Retention time t_R (min)	Pyrolysis product
1	10.79	Hexamethylcyclotrisiloxane
2	18.21	Octamethylcyclotetrasiloxane
3	23.84	Decamethylcyclopentasiloxane
4	27.85	Dodecamethylcyclohexasiloxane
5	30.68	Tetradecamethylcycloheptasiloxane
6	32.94	Hexadecamethylcyclooctasiloxane
7	34.79	Octadecamethylcyclononasiloxane
8	36.44	Eicosamethylcyclodecasiloxane
9	38.05	Docosamethylcycloundecasiloxane
10	39.70	Tetracosamethylcyclododecasiloxane
11	41.57	Hexacosamethylcyclotridecasiloxane
12	43.86	Octacosamethylcyclotetradecasiloxane
13	46.82	Triacontamethylcyclopentadecasiloxane
14	50.74	Dotriacontamethylcyclohexadecasiloxane
15	55.98	Tetratriacontamethylcycloheptadecasiloxane
16	63.01	Hexatriacontamethylcyclooctadecasiloxane
17	72.48	Octatriacontamethylcyclononadecasiloxane
18	85.24	Tetracontamethylcycloeicosasiloxane

Notes: Apparatus 2, GC column 3, GC conditions 3.

of silicon rubber (PDMS) pyrolyzed at 700°C. On the pyrogram, the peaks of cyclic oligodimethylsiloxanes of the structure $[(CH_3)_2SiO]_n$ ($n = 3$ to $n = 20$) were observed. The yield of each cyclic oligodimethylsiloxane in pyrogram (Fig. 4.41) decreases with an increase in MW from 222 ($n = 3$) to 1480 ($n = 20$).

4.6.9. Analysis of rubber blends

The use of blends of elastomers is almost as old as the synthetic rubber industry, and generally stems from an understandable desire to combine the best features — technical or economic — of two elastomers [100]. Blending or mixing of elastomers has become an increasingly important area of research activity. The processes are carried out for three main reasons: improvement of the technical properties of the original elastomer, achievement of better processing behavior, and lowering of compound cost. Elastomer blends are used for many reasons such as lowering the compound cost, for ease of fabrication, and to improve the performance of the industrial rubber. NR and SBR have been blended for a long time for these reasons. Polymer blends can be prepared by several methods like latex blending, mechanical blending, and solution blending. The elastomer phase is often a blend of different elastomers. However, there are technological problems arising from some types of mutual incompatibility that can exist between dissimilar elastomers. Three main types of incompatibility have generally been noted: the thermodynamic incompatibility, incompatibility due to viscosity mismatch, and incompatibility due to the cure rate mismatch [101]. All elastomers have deficiencies in one or more properties and blending is a way of obtaining optimum all-around performance. Compounds with good properties also need to be capable of factory processing without difficulty and of providing uniformity in behavior [101]. Mixtures and blends occur at different hierarchical scales in the material range employed in the rubber industry. Composite products such as tires, hoses, belts, and air springs are composed of metal wire, textile cord, and elastomeric compounds, which form a rubber matrix. The rubber matrix itself is a mixture of elastomer, filler, and plasticizer.

The examples described in the next sections show the pyrolysis–GC/MS chromatograms with the identification of the pyrolysis products of some rubber blends at 700°C.

4.6.9.1. *Characterization of an SBR/BR blend*

Commonly used rubbers in the automotive industry are NR (polyisoprene), synthetic polyisoprene (IR), polybutadiene (BR), styrene–butadiene copolymers (SBR), nitrile rubber (NBR), and the blends of these elastomers [102]. Elastomer blends can be classified into two main categories: miscible and immiscible blends. Both have found useful applications in tire treads. For example, miscible SBR/BR blends are used in state-of-the-art tire treads due to the advantageous set of traction, wear, and rolling resistance [103]. Immiscible IR/NR blends are used due to the improved wet-skid resistance.

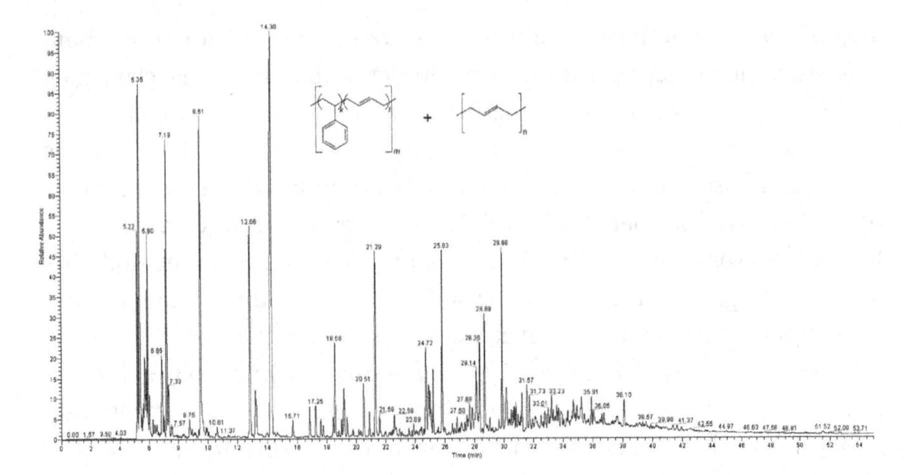

Fig. 4.42. Pyrolysis–GC/MS chromatogram of a SBR/BR blend at 700°C. Apparatus 2, GC column 3, GC conditions 3. For peak identification, see Table 4.21.

These examples illustrate that miscible and immiscible blends exhibit dissimilar advantages for tire performance. Depending on blend miscibility behavior, there are changes in tire performance [103]. Figure 4.42 shows the Py–GC/MS chromatogram of an SBR/BR blend pyrolyzed at 700°C. The pyrolysis products of the SBR/BR blend at 700°C are summarized in Table 4.21.

4.6.9.2. *Characterization of an NR/SBR blend*

The blends of NR either with polybutadiene rubber (BR) or with SBR are also widely used in tread compounds. By blending of SBR with NR, the mechanical property of the former could be improved. In the work of George *et al.* [104] the effect of the SBR/NR blend composition on morphology, transport behavior, and mechanical and dynamic mechanical properties were investigated. The thermal degradation of NR/SBR blends has been investigated by thermographic methods with reference to the effects of various parameters like blend ratio, vulcanization, and vulcanizing systems [105]. The authors stated that the degradation of SBR takes place at a higher temperature than NR. The main degradation products of NR were identified as isoprene and dipentene (isoprene dimer). SBR has

Table 4.21. Pyrolysis products of an SBR/BR blend at 700°C.

Retention time t_R (min)	Pyrolysis product
5.22	Propylene
5.35	1-Butene
5.41	2-Butene
5.50	3-Methyl-1-butene
5.63	2-Pentene
5.71	1,2-Pentadiene
5.80	1,3-Pentadiene
5.90	1,3-Cyclopentadiene
6.01	1-Cyclopentene
6.85	1,4-Cyclohexadiene
6.90	5-Methyl-1,3-cyclopentadiene
7.02	2,4-Hexadiene
7.19	Benzene
7.33	1,4-Cyclohexadiene
8.75	1-Methyl-1,4-cyclohexadiene
9.51	Toluene
12.86	Ethylbenzene
13.22	*p*-Xylene
13.27	Xylene isomer
14.30	Styrene
15.71	Cumene
16.84	Allylbenzene
17.25	*n*-Propylbenzene
17.59	1-Ethyl-3-methylbenzene
17.76	1-Ethyl-2-methylbenzene
18.44	1-Ethyl-4-methylbenzene
18.56	α-Methylstyrene
19.00	*o*-Methylstyrene
19.18	*p*-Methylstyrene
20.16	*o*-Cymene
20.51	2-Propenylbenzene

(*Continued*)

Table 4.21. (*Continued*)

Retention time t_R (min)	Pyrolysis product
20.89	Indane
21.29	Indene
24.72	1,2-Dihydronaphthalene
24.91	1,4-Dihydronaphthalene
25.83	Naphthalene
26.82	Benzothiazole
28.36	2-Methylnaphthalene
28.69	1-Methylnaphthalene
29.88	Biphenyl
31.24	Acenaphthylene
31.57	4-Methyl-1,1′-biphenyl
35.91	Phenanthrene
38.10	2-Phenylnaphthalene

Notes: Apparatus 2, GC column 3, GC conditions 3.

decomposed to the large number of low-MW compounds, like 4-phenylcyclohexene, 4-vinylcyclohexene (4-ethenylcyclohexene, butadiene dimer), methyl benzene (toluene), ethyl benzene, styrene, and α-methylstyrene. However, blends show a higher degradation temperature than NR. As the SBR content in blends increases, the degradation temperature increases, indicating increased thermal stability of the blends [105].

Figure 4.43 shows the pyrolysis–GC/MS chromatogram of an NR/SBR blend at 700°C. The pyrolysis products with the origin of the pyrolyzed part of the blend are shown in Table 4.22. Benzothiazole identified at the retention time of $t_R = 16.65$ min has been formed by the thermal degradation of 2-mercaptobenzothiazole. 2-Mercaptobenzothiazole is used as an accelerator for the vulcanization of rubber and as an antioxidant [12, 47].

4.6.9.3. *Characterization of an CR/SBR blend*

Polychloroprene/SBR (CR/SBR) blend is used for calendered sheeting. Laminated products of CR/SBR rubber sponge with various fabrics are

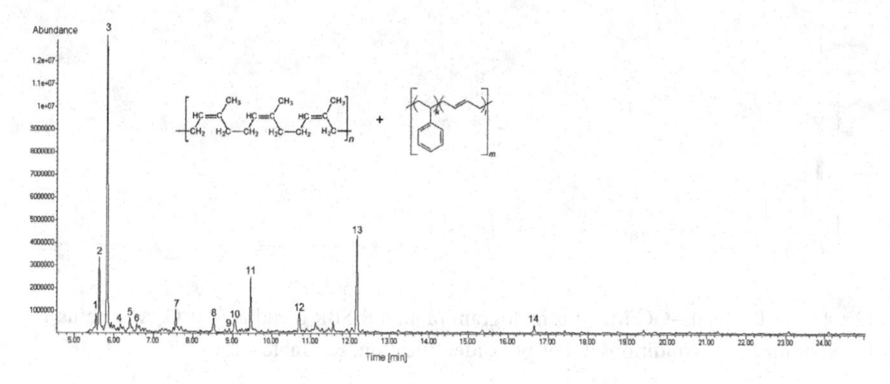

Fig. 4.43. Pyrolysis–GC/MS chromatogram of an NR/SBR blend at 700°C. Apparatus 1, GC column 1, GC conditions 2. For peak identification, see Table 4.22.

Table 4.22. Pyrolysis products of an NR/SBR blend at 700°C.

Peak No.	Retention time t_R (min)	Pyrolysis product	Origin
1	5.55	Propylene	SBR, NR
2	5.63	1,3-Butadiene	SBR
3	5.83	2-Methyl-1,3-butadiene (Isoprene)	NR
4	6.17	3-Methyl-2-pentene	NR
5	6.41	5-Methyl-1,3-cyclopentadiene	SBR
6	6.58	Benzene	SBR
7	7.57	Toluene	SBR
8	8.53	4-Ethenylcyclohexene (1,3-Butadiene dimer)	SBR
9	8.92	Ethylbenzene	SBR
10	9.06	*p*-Xylene	SBR
11	9.46	Styrene	SBR
12	10.71	1-Methyl-4-(1-methylethenyl)-cyclohexene	NR
13	12.17	Dipentene (Limonene, isoprene dimer)	NR
14	16.65	Benzothiazole	Vulcanization accelerator

Notes: Apparatus 1, GC column 1, GC conditions 2.

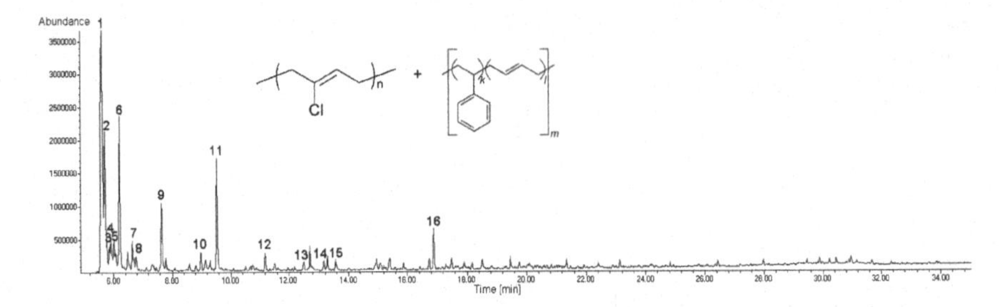

Fig. 4.44. Pyrolysis–GC/MS chromatogram of an CR/SBR blend at 700°C. Apparatus 1, GC column 1, GC conditions 2. For peak identification, see Table 4.23.

suitable materials for shoes, fishing boots, horse blankets, medical care, and sport protection accessories [106]. CR/SBR blends are used for gaskets because of its good resistance to many types of oils and greases. They also have good resistance to ozone, sunlight, and atmospheric ageing. The operating temperature of CR/SBR blends ranges from –20°C to +70°C. Figure 4.44 shows the pyrolysis–GC/MS chromatogram of an CR/SBR blend at 700°C. The identified degradation products of the investigated CR/SBR blend are summarized in Table 4.23. The mass spectra of the characteristic pyrolysis products of polychloroprene (CR), like chloroprene monomer and chloroprene dimer, are shown in Fig. 4.45.

Other rubber blend consisting of NR and polybutadiene rubber (BR) will be shown in Section 7.3.3.

4.7. Analysis of Polyacrylates

4.7.1. Characterization of poly(alkyl acrylates)

PMMA, the most popular poly(alkyl acrylate), known as *Plexiglas*® was invented by Otto Röhm in 1933 in Germany. PMMA is produced by free-radical polymerization of methyl methacrylate (MMA), according to the Eq. (4.1):

$$n\,\mathrm{CH_2{=}C(CH_3)(COOCH_3)} \rightarrow [\mathrm{-CH_2{-}C(CH_3)(COOCH_3)-}]_n, \qquad (4.1)$$

MMA PMMA

Equation (4.1) describes the polymerization of MMA to form poly(methyl methacrylate) (PMMA).

Table 4.23. Pyrolysis products of an CR/SBR blend at 700°C.

Peak No.	Retention time t_R (min)	Pyrolysis product	Origin
1	5.61	Carbon dioxide	SBR
2	5.71	1,3-Butadiene	SBR
3	5.86	3-Methyl-1-butene	SBR
4	5.90	1,4-Pentadiene	SBR
5	6.01	1,3-Cyclopentadiene	SBR
6	6.21	2-Chloro-1,3-butadiene (Chloroprene monomer)	CR
7	6.49	1,4-Cyclohexadiene	SBR
8	6.65	Benzene	SBR
9	7.65	Toluene	SBR
10	8.99	Ethylbenzene	SBR
11	9.53	Styrene	SBR
12	11.20	α-Methylstyrene	SBR
13	12.71	Indene	SBR
14	13.17	2-Chlorostyrene	CR
15	13.57	4-Chloro-1,2-dimethylbenzene	CR
16	16.89	1-Chloro-4-(1-chloroethenyl)-cyclohexene (Chloroprene dimer)	CR

Notes: Apparatus 1, GC column 1, GC conditions 2.

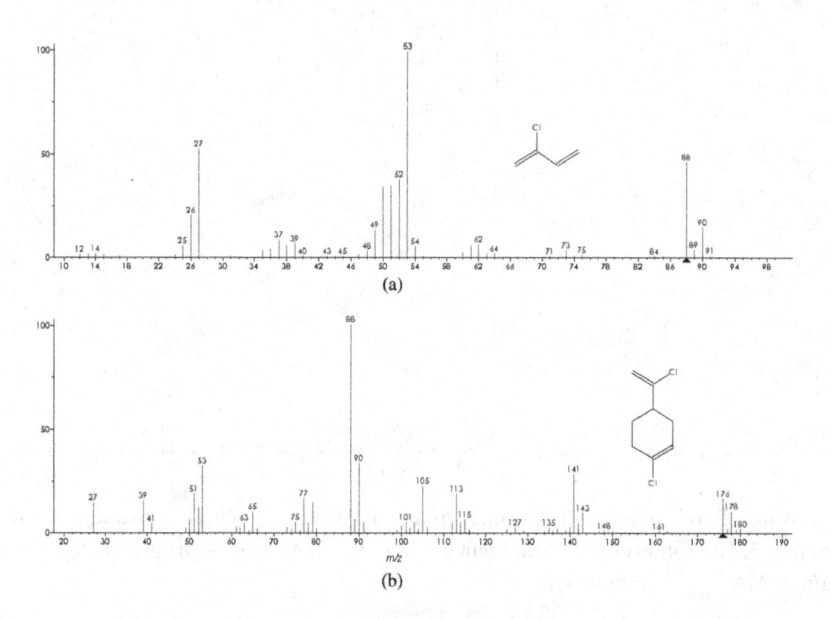

Fig. 4.45. Mass spectrum of (a) 2-chloro-1,3-butadiene (chloroprene monomer), and (b) 1-chloro-4-(1-chloroethenyl)-cyclohexene (chloroprene dimer).

The advantages of PMMA are glass-clear transparency and high impact strength. The applications of *Plexiglas*® range extend from light-transmitting and weather-resistant glazing in private buildings and greenhouses to industrial buildings, exhibition stands, and protective noise barriers. PMMA is used as glazing in military aircrafts and helicopters. *Plexiglas*® is suitable for many optic/electronic applications such as cover glazing for displays or TVs and rear projection screens. PMMA has found use in the production of bath, health, and heat therapy equipment as well as for working and furniture surfaces in kitchens and laboratories [1].

Poly(alkyl acrylate)s depolymerize (retropolymerize) upon heating to their constituent monomers [1]. The thermal decomposition of PMMA yields MMA (87.5%) and methyl acrylate (MA) (2.5%) [1]. Figure 4.46 shows the pyrolysis–GC/MS chromatogram of PMMA obtained after pyrolysis at 700°C.

Because of the structure of methacrylate polymers, the favored degradation is essentially a reversion to the monomer [13]. For the most part monomer production is unaffected by the R group, so that PMMA will revert to MMA, poly(ethyl methacrylate) will produce ethyl methacrylate

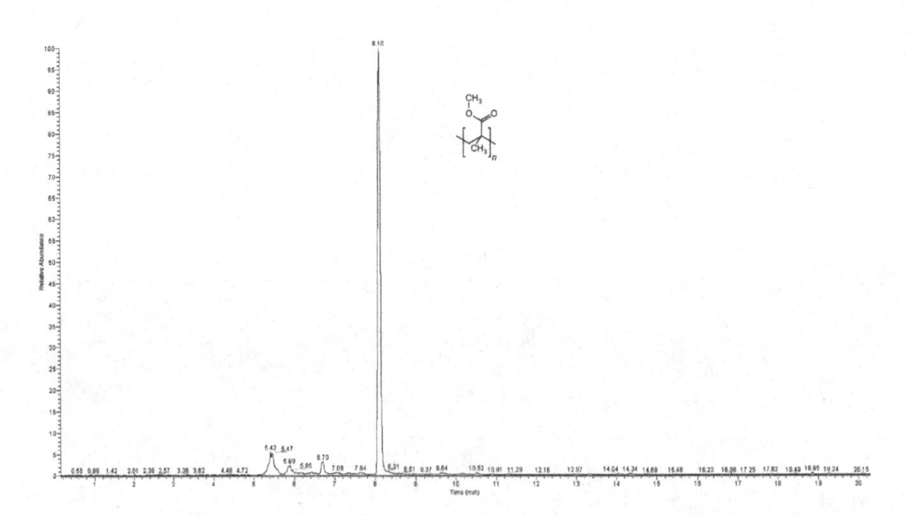

Fig. 4.46. Pyrolysis–GC/MS chromatogram of PMMA at 700°C. Apparatus 2, GC column 3, GC conditions 3. Peak identification: t_R = 5.42 min — propylene, t_R = 6.70 min — MA, t_R = 8.10 min — MMA.

(EMA), etc. [13]. This proceeds in copolymers as well, with the production of both monomers in roughly the original polymerization ratio. A pyrogram of a copolymer of two or more acrylate monomers or a blend of two polyacrylates would contain peaks for each of the monomers in the polymer [107]. For example, the pyrogram of the homopolymer mixture (blend) of 30% (m/m) PMMA and 70% (m/m) poly(*n*-butyl acrylate) (PBA) shows the peaks of the monomers MMA and *n*-butyl acrylate [107]. Homogeneous polymer mixtures of PBA/PMAA are used in the optics industry for production of contact lenses.

The thermal decomposition behavior of a series of dimethacrylate monomers and their crosslinked polymers was investigated in the work of Andrzejewska *et al.* [108] by using Curie-point pyrolysis/GC. The formation of the corresponding monomers and methacrylic acid (MAA) was monitored to evaluate the relative importance of the two processes, depolymerization and ester group decomposition. As expected, the predominance of either of these processes depends very much upon the structure of the ester. In the case of a sulfur-containing polymer, the predominant process of pyrolysis is ester group decomposition and MAA formation due to the presence of a weak C–S bond and easily abstractable H atoms [108].

Characteristic for the pyrolysis of poly(acrylic acid) (PAA) is the formation of carbon dioxide (CO_2).

The polymers poly(vinylidene fluoride-*co*-hexafluoropropylene) (PVDF–HFP) and poly(*n*-butyl methacrylate) (PBMA) are employed in manufacturing the XIENCE family of coronary stents. PBMA serves as a primer and adheres to both the stent and the drug coating. PVDF–HFP is employed in the drug matrix layer to hold the drug everolimus on the stent and control its release. Chemical stability of the polymers of XIENCE stents in the in vivo environment was evaluated by pyrolysis–GC with MS in the work of Kamberi *et al.* [109]. For this evaluation, XIENCE stents explanted from porcine coronary arteries and from human coronary artery specimens at autopsy after 2–4 and 5–7 years of implantation, respectively, were compared to freshly manufactured XIENCE stents (controls). The comparison of pyrograms of explanted stent samples and controls showed identical fragmentation fingerprints of polymers, indicating that PVDF–HFP and PBMA maintained their chemical integrity after multiple years of XIENCE coronary stent

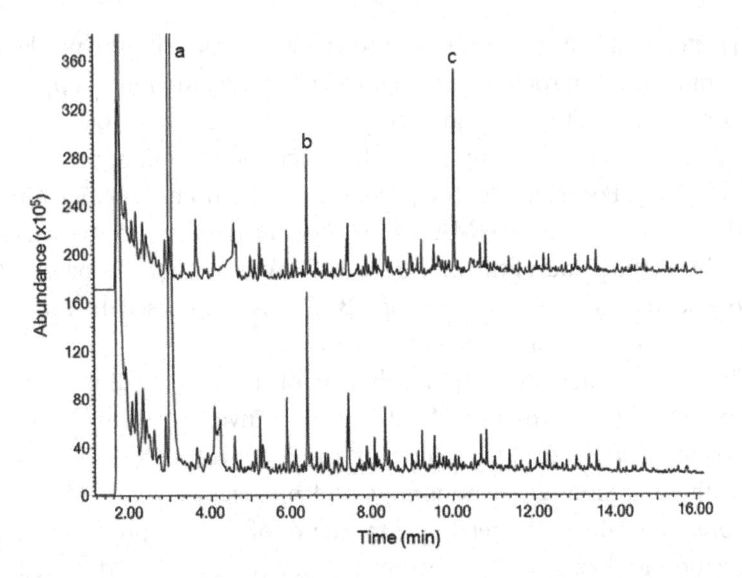

Fig. 4.47. Pyrograms of freshly manufactured (top) and explanted (autopsy) (bottom) XIENCE coronary stents. (a) MMA (monomer of PMMA), the embedding material; (b) monomer of PBMA; (c) everolimus [109].

implantation. The findings of the study demonstrate the chemical stability of PVDF–HFP and PBMA polymers of the XIENCE family of coronary stents in the in vivo environment and constitute a further proof of the suitability of PVDF–HFP as a drug carrier for the drug eluting stent applications [109]. Figure 4.47 shows the representative pyrograms obtained by Kamberi *et al.* [109] of freshly manufactured and explanted (autopsy) XIENCE coronary stents.

4.7.2. Polyacrylates in the dentistry

A number of dental filling materials are presently available for tooth restorations. The four main groups of these materials, which dentists have used for about 40 years, are conventional glass–ionomer cements, resin-based composites, resin-modified glass–ionomer cements and polyacid-modified resinous composites [110]. Light-curing glass–ionomer cements contain poly(acrylic acid) (PAA), chemically and/or photo-curing monomers (multifunctional methacrylates, like triethylene glycol dimethacrylate or 2-hydroxyethyl methacrylate), an ion-leaching glass and additives (initiators, inhibitors, stabilizers, and others) [110]. Resin-modified glass–ionomer cements are now widely used in dentistry as direct filling materials, liners, bases, luting cements, and fissure sealants [111]. These materials mainly consist of polymer matrix and glass–ionomer parts. The polymer matrix is based on a monomer system and different multifunctional methacrylates with additives [111]. Methacrylic monomers, like bisphenol A glycidyl methacrylate (Bis-GMA), urethane dimethacrylate (UDMA), triethylene glycol dimethacrylate (TEGDMA), and 2-hydroxyethyl methacrylate (HEMA) are the main components of resin-based dental filling materials. The presence of additives such as initiators, activators, inhibitors, and plasticizers in uncured dental material mixture is necessary [111].

Figure 4.48 shows the TIC Py–GC/MS chromatogram of commercially light-curing dental filling material pyrolyzed at 550°C. The pyrolysis products identified by using mass spectra library *NIST 05* are summarized in Table 4.24. The carbon dioxide (t_R = 5.85 min) identified in pyrolyzate is formed from poly(acrylic acid) (PAA). The identified substances, like HEMA (t_R = 13.65 min), EGDMA (t_R = 19.48 min), and

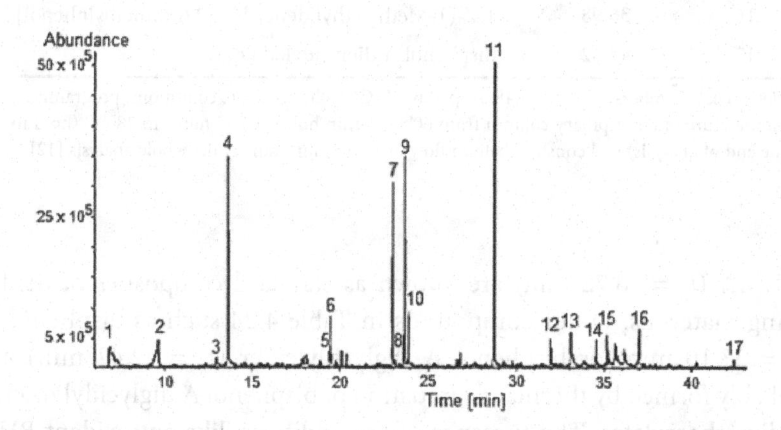

Fig. 4.48. Pyrolysis–GC/MS chromatogram of commercially light-curing dental filling material pyrolyzed at 550°C. For peak identification, see Table 4.24. Apparatus 1, GC column 2, GC conditions: programmed temperature of the capillary column from 60°C (1-min hold) at 7°C/min to 280°C (hold to the end of analysis) and constant helium flow of 1 cm³/min during the whole analysis [12].

Table 4.24. Pyrolysis products of commercially light-curing dental filling material at 550°C.

Peak No.	Retention time t_R (min)	Pyrolysis products of the dental filling material at 550°C
1	5.85	Carbon dioxide
2	9.62	Methacrylic acid (MAA)
3	12.96	Phenol
4	13.65	2-Hydroxyethyl methacrylate (HEMA)
5	19.40	4-Isopropenylphenol
6	19.48	Ethylene glycol dimethacrylate (EGDMA)
7	23.00	Not identified
8	23.17	2,6-Bis-(1,1-dimethylethyl)-4-methylphenol (BHT)
9	23.65	Not identified
10	23.89	Triethylene glycol (TEG)
11	28.72	Triethylene glycol dimethacrylate (TEDMA)
12	31.95	Drometrizol (Tinuvin-P, UV-absorber)
13	33.10	4,4′-Dihydroxy-2,2-diphenylpropane (Bisphenol A)
14	34.55	Triphenylantimony
15	35.25	Tetraethylene glycol dimethacrylate
16	36.98	4,4′-(1-Methylethylidene)-bis-[2,6-dimethylphenol]
17	42.42	Bisphenol A diglycidylether

Notes: Peak numbers as in Fig. 4.48. Apparatus 1, GC column 2, GC conditions: programmed temperature of the capillary column from 60°C (1-min hold) at 7°C min^{-1} to 280°C (hold to the end of analysis) and constant helium flow of 1 cm^3 min^{-1} during the whole analysis [12].

TEDMA (t_R = 28.72 min) are known as standard composites of dental filling materials. Other compounds in Table 4.24 such as bisphenol A (t_R = 33.10 min) or bisphenol A diglycidylether (t_R = 42.42 min) are probably formed by thermal degradation of bisphenol A diglycidyl mono- or dimethacrylates. The presence of the additives, like antioxidant BHT [2,6-bis-(1,1-dimethylethyl)-4-methylphenol, t_R = 23.17 min] or the UV-absorber drometrizol (t_R = 31.95 min), was also confirmed. The triphenylantimony (t_R = 34.55 min) identified in pyrolyzate is used as catalyst in the UV-induced polymerization.

4.7.3. Characterization of copolymers of methacrylic acid with poly(ethylene glycol) methyl ether methacrylate macromonomers

Amphiphilic graft copolymers with a hydrophobic backbone and hydrophilic grafts can be synthesized by copolymerization of macromonomers [112]. Macromonomers are typically produced by reaction of poly(ethylene glycol) (PEG) or PEG monomethyl ether (PEG-ME) with (meth) acrylic acid or (meth) acryloyl chloride. Monoacrylates or methacrylates of poly(ethylene glycol) or its monoalkyl or monoaryl ethers can be polymerized (or copolymerized) with other monomers to yield graft copolymers with a (meth) acrylate backbone and PEG grafts. The corresponding di(meth)acrylates are crosslinkers for the production of hydrogels [112].

Hydrogels are crosslinked (co) polymers with a 3D network arrangement in which individual hydrophilic (co) polymer chains are connected by physical or chemical bonds. Depending on the hydrophilicity, they can absorb water up to thousands of times of their dry weight and form chemically stable or biodegradable gels [112]. Hydrogels are defined as smart, active, intelligent, programmed, and shaped memory polymeric materials. Hydrogels based on natural or synthetic polymers have been of great interest regarding cell encapsulation. For the past decade, they have become especially attractive as matrices for regenerating and repairing a wide variety of soft tissues and organs [112]. Hydrogels are an appropriate biomaterial in implantology, e.g. in the corneal implants. Copolymers based on poly(ethylene glycol) methyl ether methacrylate (ME-PEG-MA) macromonomers have been of attracting interest due to their unique properties, such as their highly different natures depending on the sizes and length of the PEG in the branch chains. Due to the fact that ME-PEG-MA macromonomers show a good solubility in water and in most organic solvents, they are the ideal candidates for applications in biomaterials, in morphology of micelles, in ion conductivity studies, and for preparation of thermoresponsive copolymers [112]. The characterization of the macromonomers is important for many applications.

In our laboratory, poly(ethylene glycol) monomethyl ether methacrylate (ME-PEG350-MA) macromonomers (Fig. 4.49(a)) were analyzed by GC/MS. Their polymers and copolymers with MAA (Fig. 4.49(b)) were

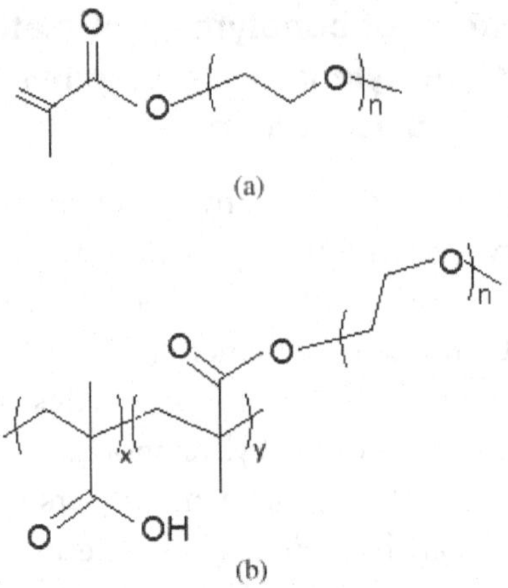

(a)

(b)

Fig. 4.49. Chemical structures of the investigated (a) poly(ethylene glycol) methyl ether methacrylate macromonomers, and (b) copolymers of MAA with poly(ethylene glycol) methyl ether methacrylate macromonomers.

characterized by analytical pyrolysis–GC/MS at 550°C. Figure 4.50 shows the obtained GC/MS chromatograms of poly(ethylene glycol) monomethyl ether methacrylate (ME-PEG350-MA) macromonomers and the chromatograms of the pyrolysates of poly(ME-PEG350-MA) and poly-(MMA-*co*-ME-PEG350-MA) at 550°C, respectively. The separated substances were identified by using mass spectra library *NIST 05* and by calculation of increments of the retention indices in programmed GC (I_p). The results are summarized in Tables 4.25 and 4.26, respectively. The obtained analytical results were then used to calculate the average degrees of ethoxylation of the macromonomers and copolymers [112]. The calculated values are also shown in Tables 4.25 and 4.26.

4.8. Analysis of Polyesters

4.8.1. Characterization of poly(ethylene terephthalate) (PET)

Typical terephthalate polyesters are poly(ethylene terephthalate) (PET) and poly(butylene terephthalate) (PBT). PET is the most important thermoplastic polyester. The polyester is produced in the form of fibers, films,

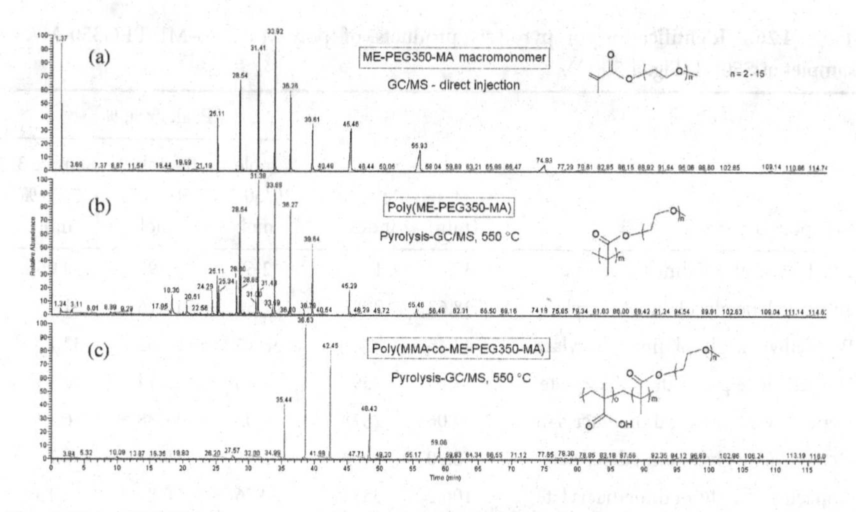

Fig. 4.50. GC/MS chromatograms of (a) ME-PEG350-MA macromonomer, (b) pyrolysate of poly(ME-PEG350-MA) at 550°C, (c) pyrolysate of poly(MMA-*co*-ME-PEG350-MA) at 550°C. Apparatus 2, GC column 3, GC conditions 3. For peak identification, see Tables 4.25 and 4.26.

Table 4.25. Identification of ME-PEG350-MA macromonomer by direct GC/MS analysis (Fig. 4.50(a)).

Compound	t_R (min)	Retention index I_p	Peak area %
Diethylene glycol monomethyl ether methacrylate	19.99	1215	1.29
Triethylene glycol monomethyl ether methacrylate	25.11	1548	2.34
Tetraethylene glycol monomethyl ether methacrylate	28.54	1823	6.66
Pentaethylene glycol monomethyl ether methacrylate	31.41	2120	16.91
Hexaethylene glycol monomethyl ether methacrylate	33.92	2389	20.28
Heptaethylene glycol monomethyl ether methacrylate	36.26	2637	17.27
Octaethylene glycol monomethyl ether methacrylate	39.61	2899	11.08
Nonaethylene glycol monomethyl ether methacrylate	45.46	3193	9.04
Decaethylene glycol monomethyl ether methacrylate	55.93	3423	7.09
Undecaethylene glycol monomethyl ether methacrylate	74.83	3728	3.74
Dodecaethylene glycol monomethyl ether methacrylate	109.14	4001	3.20
Others			1.01
Average degree of ethoxylation α ($n = 4$)			7.5 (± 0.04)

Notes: Apparatus 2, GC column 3, GC conditions 3.

Table 4.26. Identification of pyrolysis products of poly(MMA-*co*-ME-PEG350-MA) samples at 550°C (Fig. 4.50(c)).

			Peak area, %		
Compound	t_R (min)	Retention index I_p	Sample 1 70/30 % mol	Sample 2 50/50 % mol	Sample 3 30/70 % mol
Triethylene glycol dimethacrylate	35.44	1821	2.69	5.95	4.06
Tetraethylene glycol dimethacrylate	38.63	2085	27.88	16.16	31.48
Pentaethylene glycol dimethacrylate	42.45	2361	38.15	21.65	33.27
Hexaethylene glycol dimethacrylate	48.43	2597	21.86	25.14	20.93
Heptaethylene glycol dimethacrylate	59.06	2873	7.05	17.38	6.13
Octaethylene glycol dimethacrylate	78.30	3156	1.69	9.45	2.76
Nonaethylene glycol dimethacrylate	106.2	3411	0.56	0.84	1.15
Others			0.12	3.43	0.22
Average degree of ethoxylation α ($n = 4$)			4.9 (± 0.05)	5.4 (± 0.06)	4.7 (± 0.03)

Notes: Apparatus 2, GC column 3, GC conditions 3.

or granules. The crystalline form of PET is produced by polycondensation of ethyl terephthalate, obtained from transesterification of dimethyl terephthalate (DMT) and ethylene glycol. The properties that particularly characterize PET are very good chemical resistance, high mechanical properties, excellent transparency and gloss, and high barrier properties, especially for oxygen and carbon dioxide [1, 96]. This combination of properties renders PET particularly attractive for food and beverage packaging. A large proportion of this polymer is used for the production of gastight bottles for carbonated beverages.

Figure 4.51 shows the results obtained by pyrolysis of PET at 700°C followed by GC/MS. As can be seen from Fig. 4.51 the main degradation products of PET are carbon dioxide (peak 1, t_R = 4.71 min), benzene (peak 2, t_R = 5.70 min), toluene (peak 3, t_R = 6.64 min), *p*-xylene (peak 4, t_R = 8.10 min), acetophenone (peak 5, t_R = 11.97 min), benzoic acid (peak 6, t_R = 14.27 min), biphenyl (peak 7, t_R = 18.59 min), *p*-diacetylbenzene (peak 8, t_R = 19.79 min), benzophenone (peak 9, t_R = 23.12 min), and fluorenone (peak 10, t_R = 25.17 min).

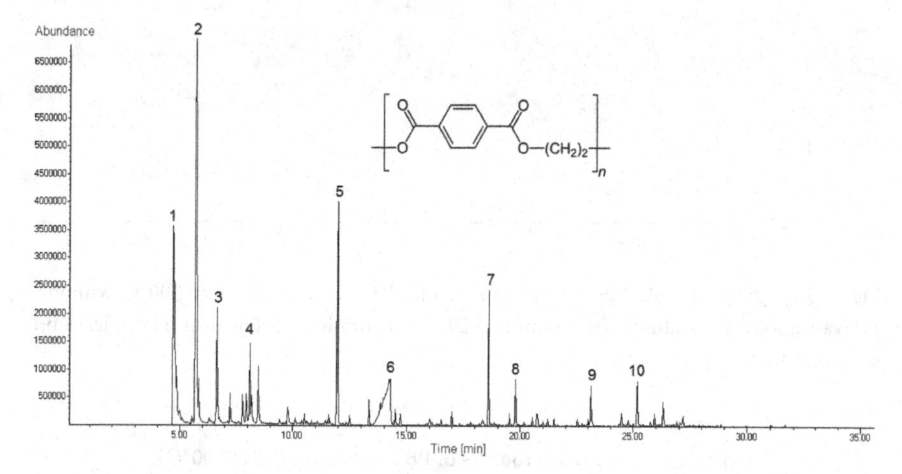

Fig. 4.51. Pyrolysis–GC/MS chromatogram of commercial PET at 700°C without derivatization. Apparatus 1, GC column 1, GC conditions 2. For peak identification, see text.

4.8.2. Characterization of poly(butylene terephthalate) (PBT)

PBT is the saturated polyester obtained from butanediol and terephthalic acid or DMT. PBT is used in the plastics industry as the basis for high-performance compounds or blends, e.g. with ASA (acrylonitrile-styrene-acrylate copolymer), PET, or PC. The large PBT suppliers are generally backward integrated to the polymer. PBT materials are characterized by high-dimensional stability, high stiffness, and good heat resistance. By incorporating fillers, reinforcing materials, and additives during compounding, material properties can be tailored to requirements [113]. In this way, materials suitable for many different applications can be produced. They are processed mainly by injection molding, but unrein forced PBT is also used in special extrusion and fiber spinning processes. This injection-molding-friendly, semi-crystalline polymer is used, for example, to produce electronics housings for auto manufacture and communications electronics. Apart from automobiles, PBT can also be used in many other applications — sometimes in combination with other plastics. Examples range from shower heads and artificial grass to insulin pens and fiber-optic cables [113].

Figure 4.52 shows the pyrolysis–GC/MS chromatogram of PBT *Vestodur 2000* (BASF, Ludwigshafen, Germany) at 700°C. The obtained pyrolysis products of PBT are summarized in Table 4.27.

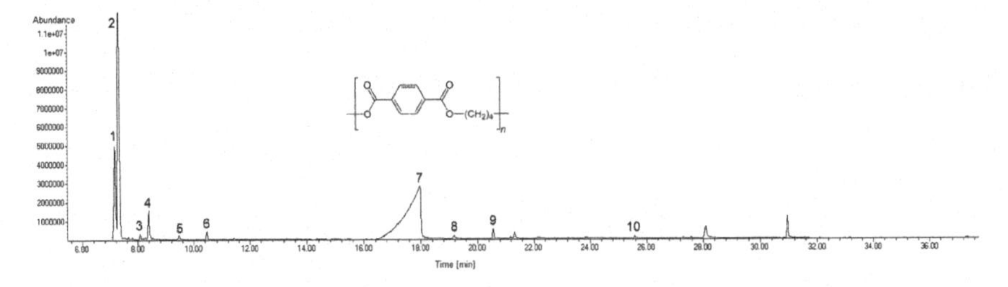

Fig. 4.52. Pyrolysis–GC/MS chromatogram of PBT *Vestodur 2000* at 700°C without derivatization. Apparatus 1, GC column 1 UI, GC conditions 2. For peak identification, see Table 4.27.

Table 4.27. Pyrolysis products of PBT *Vestodur 2000* at 700°C.

Peak No.	Retention time t_R (min)	Pyrolysis product
1	7.15	Carbon dioxide
2	7.29	1,3-Butadiene
3	8.08	Tetrahydrofuran
4	8.39	Benzene
5	9.47	Toluene
6	10.45	4-Ethenylcyclohexene
7	17.97	Benzoic acid
8	19.17	Toluic acid
9	21.32	Biphenyl
10	25.59	Benzophenone

Notes: Apparatus 1, GC column 1 UI, GC conditions 2.

During the pyrolysis–GC/MS of PET or PBT some peaks cannot be observed in the pyrograms, which correspond to monomers and/or short oligomers having one or more polar functional group, like carboxyl group (-COOH). These types of compounds are very often retained inside the chromatographic capillary column because of their high polarity and low volatility [114]. In order to overcome this disadvantage, a combination of pyrolysis of PET or PBT with *in situ* methylation (derivatization), by using tetramethylammonium hydroxide (TMAH) or tetramethylammonium acetate (TMAAc) reagents, was introduced [114, 115]. Through the

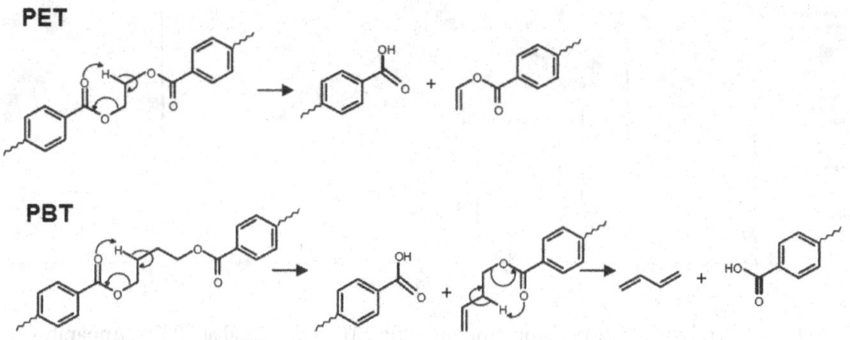

Fig. 4.53. Intramolecular decomposition reactions of terephthalate polyesters PET and PBT proposed by Kawai *et al.* [115]. Scheme reprinted from Ref. [115] with permission from Elsevier.

thermally assisted chemolysis with TMAH or TMAAc, polymer chains were decomposed selectively at carbonate and/or ester linkages to yield quantitatively the methyl derivatives of the components reflecting not only the main chain but also the branching and crosslinking structures [115]. Figure 4.53 shows the scheme of the intramolecular decomposition reactions of the terephthalate polyesters PET and PBT proposed by Kawai *et al.* [115].

4.9. Analysis of Polycarbonates

Polycarbonate (PC) was discovered by Hermann Schnell in 1953 at Bayer AG (Germany). PCs are an unusual and extremely useful class of high heat polymers known for their toughness and clarity. The vast majority of PCs is based on bisphenol A (BPA) and sold under the trade names *Lexan* (GE), *Makrolon* (Bayer), *Caliber* (Dow), *Panlite* (Teijin), and *Iupilon* (Mitsubishi) [116]. BPA PCs have glass transition temperatures (T_g) in the range of 140–155°C and are widely regarded for their optical clarity and exceptional impact resistance and ductility at or below room temperature. Other properties such as modulus, dielectric strength, and tensile strength are comparable to other amorphous thermoplastics at similar temperatures below their respective T_gs. PCs are prepared commercially by two completely different processes: Schotten–Baumann reaction of phosgene and an aromatic diol in

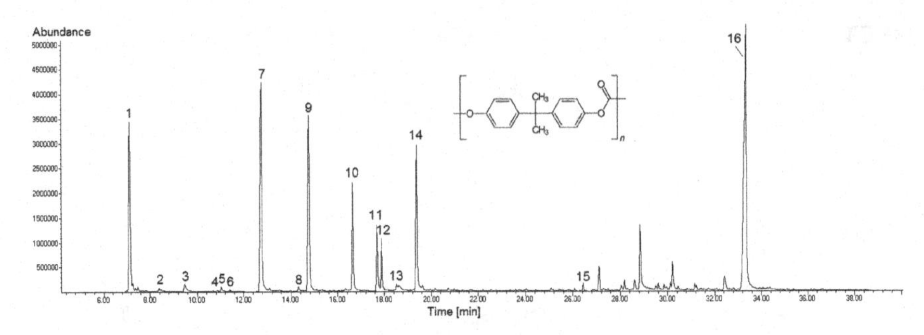

Fig. 4.54. Pyrolysis–GC/MS chromatogram of PC *Makrolon 2800* at 700°C. Apparatus 1, GC column 1 UI, GC conditions 2. For peak identification, see Table 4.28.

an amine-catalyzed interfacial condensation reaction, or via base-catalyzed transesterification of a bisphenol with a monomeric carbonate such as diphenyl carbonate. Many important products are also based on PC in blends with other materials, copolymers, branched resins, flame-retardant compositions, foams, and other materials [116]. PC polymers are used to produce a variety of materials and are particularly useful when impact resistance and/or transparency are a product requirement (e.g. in bullet-proof glass). PC is commonly used for plastic lenses in eyewear, in medical devices, automotive components, protective gear, greenhouses, digital disks (CDs, DVDs, and Blu-ray) and exterior lighting fixtures. PCs also have very good heat resistance and can be combined with flame-retardant materials without significant material degradation. PC plastics are engineering plastics. They are typically used for more capable, robust materials such as in impact resistant "glass-like" surfaces [117].

Figure 4.54 shows the pyrolysis–GC/MS chromatogram of PC *Makrolon 2800* (Bayer AG, Leverkusen, Germany) based on bisphenol A at 700°C. The main pyrolysis product of *Makrolon 2800* is bisphenol A (peak 16, t_R = 33.34 min). Figure 4.55 shows the mass spectrum of bisphenol A. The pyrolysis products of this PC are summarized in Table 4.28.

4.10. Analysis of Epoxy Resins

Epoxy resins are polyaddition products of epichlorohydrin (1-chloro-2,3-epoxy-propane) with materials that possess at least two reactive hydrogen

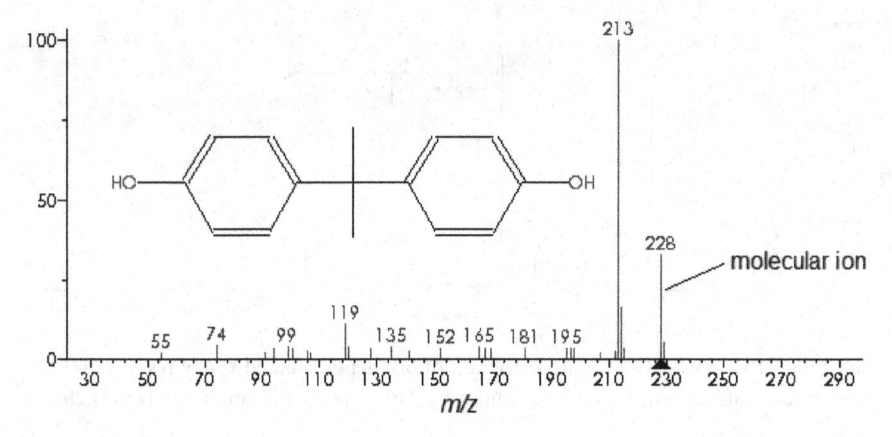

Fig. 4.55. Mass spectrum of bisphenol A identified as the main pyrolysis product of bisphenol A-based PC.

Table 4.28. Pyrolysis products of PC *Makrolon 2800* at 700°C.

Peak No.	Retention time t_R (min)	Pyrolysis product
1	7.14	Carbon dioxide
2	8.38	Benzene
3	9.47	Toluene
4	10.86	Ethylbenzene
5	11.03	*p*-Xylene
6	11.40	Styrene
7	12.76	Phenol
8	14.32	*o*-Cresol
9	14.76	*p*-Cresol
10	16.66	4-Ethylphenol
11	17.69	2,3-Dihydrobenzofuran
12	17.88	4-Isopropylphenol
13	18.51	2-Propylphenol
14	19.38	4-Isopropenylphenol
15	26.42	Diphenylcarbonate
16	33.34	4,4′-(1-Methylethylidene)bisphenol (Bisphenol A)

Notes: Apparatus 1, GC column 1 UI, GC conditions 2.

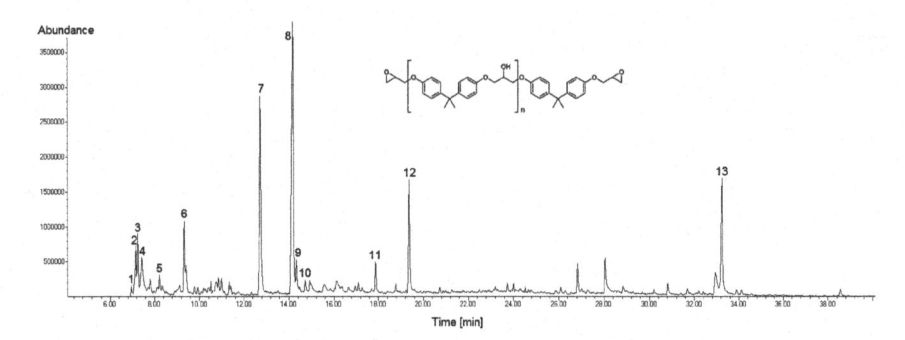

Fig. 4.56. Pyrolysis–GC/MS chromatogram of bisphenol A-based epoxy resin at 700°C. Apparatus 1, GC column 1 UI, GC conditions 2. For peak identification, see Table 4.29.

atoms. These include polyphenolic compounds, mono- and diamines, aminophenols, heterocyclic imides and amides, aliphatic diols and polyols, and dimeric fatty acids [1]. Epoxy resins derived from epichlorohydrin and bisphenol are termed *bisphenol A-based glycid-ethers resins.* The broad scope of epoxy resin applications ranges from communications through the automotive industry, aeronautics and space flight, heavy-duty corrosion protection, through ski, tennis, and sailing sports equipment, the electrical industry and wind energy up to the field of civil engineering, paints and coating, adhesives, fiber composites, and laminates [1]. Figure 4.56 shows the pyrolysis–GC/MS chromatogram of bisphenol A-based epoxy resin at 700°C. The degradation products are summarized in Table 4.29. Acetone is a minor degradation component but is an indicative pyrolysis product of the 2-hydroxypropylether segment of epoxy resin (see chemical structure in Fig. 4.56). It is not formed from other bisphenol A-based polymers such as PCs, polysulfones (PSFs) or polyethers, which decompose by pyrolysis to the same phenolic compounds as epoxy resins [1].

4.11. Analysis of Polyaryletherketones

Polyaryletherketones (PAEK) are thermoplastic polymers in which the structure of the phenylene rings are coupled through oxygen bridges (ether groups) and carbonyl (ketone) groups (see chemical structure in Fig. 4.57). The most commercially significant PAEK, polyetheretherketone (PEEK), is prepared by reacting hydroquinone (1,4-benzenediol)

Table 4.29. Pyrolysis products of bisphenol A-based epoxy resin at 700°C.

Peak No.	Retention time t_R (min)	Pyrolysis product
1	6.96	Carbon dioxide
2	7.17	Propylene
3	7.26	1-Butene
4	7.45	*n*-Butane + acetone
5	8.22	Benzene
6	9.33	Toluene
7	12.74	Benzaldehyde
8	14.18	Benzyl alcohol
9	14.32	*o*-Cresol
10	14.76	*p*-Cresol
11	17.88	4-Isopropylphenol
12	19.38	4-Isopropenylphenol
13	33.34	4,4′-(1-Methylethylidene)bisphenol (Bisphenol A)
14	38.55	2,4-Bis(1-phenylethyl)phenol

Notes: Apparatus 1, GC column 1 UI, GC conditions 2.

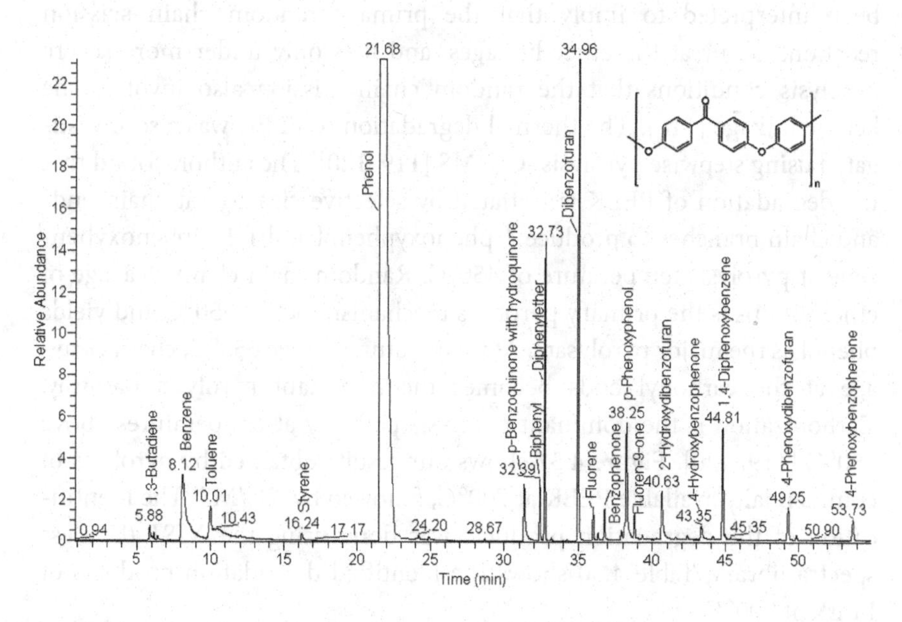

Fig. 4.57. Pyrolysis–GC/MS chromatogram of PEEK *Victrex*® at 700°C [1]. Apparatus 2, GC column 4, GC conditions 3. For peak identification, see Table 4.30.

with 4,4′-difluorobenzophenone in diphenylsulfone, in the presence of alkali-metal carbonates, under an inert atmosphere at temperatures above 300°C [1]. Other PAEK are polyetherketone (PEK), polyetherketoneketone (PEKK), polyetheretherketoneketone (PEEKK), and polyetherketoneetherketoneketone (PEKEKK). Commercial PAEK is hard and displays excellent wear and abrasion resistance. The thermal stability of PAEK is also outstanding. The melting point of PEEK is 340°C. PEEK is insoluble in all common organic solvents. It is soluble at room temperature in strongly acidic media (e.g. concentrated sulfuric acid, hydrofluoric acid, and trifluoromethanesulfonic acid) [1]. The uses of PAEK are generally connected with their outstanding high-temperature performance and chemical resistance. An important end use of PAEK is in the production of carbon fiber composite which is widely used in the aerospace and automobile industries. PEEK is used in electronics, medical and laboratory equipment, and in transportation [1].

The thermal degradation of PAEK has been studied over a range of pyrolysis conditions using both a flash pyrolysis–GC/MS and thermogravimetry/mass spectrometry (TG/MS) techniques [118]. The major thermal decomposition product was found to be phenol. The results have been interpreted to imply that the primary random chain scission reactions occur at the ether linkages, and it is only under more severe pyrolysis conditions that the random chain scission also involves the ketone linkage [118]. The thermal degradation of PEEK was also investigated using stepwise pyrolysis–GC/MS [119, 120]. The authors found that the degradation of PEEK is initiated by selective cleavage at chain ends and chain branches to produce 4-phenoxyphenol and 1,4-diphenoxybenzene at pyrolysis temperature of 450°C. Random main chain cleavage of ether groups is the primary pyrolysis mechanism below 650°C and yields phenol as the major pyrolysate. At temperatures above 650°C, chain cleavage at the carbonyl ends becomes the dominant pyrolysis pathway. Carbonization is the dominant pyrolysis pathway at temperatures above 750°C [119, 120]. Figure 4.57 shows our results obtained by pyrolysis of commercially available PEEK at 700°C, followed by GC/MS. The identification of the degradation products was done using the *NIST 05* mass spectra library. Table 4.30 shows the identified degradation products of PEEK at 700°C.

Table 4.30. Pyrolysis products of PEEK at 700°C [1].

Retention time t_R (min)	Compound
5.88	1,3-Butadiene
8.12	Benzene
10.01	Toluene
16.24	Styrene
21.68	Phenol
31.35	*p*-Benzoquinone
31.35	Hydroquinone
32.39	Biphenyl
32.73	Diphenylether
34.96	Dibenzofuran
36.00	Fluorene
36.77	Benzophenone
38.25	*p*-Phenoxyphenol
38.80	Fluoren-9-one
40.63	2-Hydroxydibenzofuran
43.35	4-Hydroxybenzophenone
44.81	1,4-Diphenoxybenzene
49.25	4-Phenoxydibenzofuran
53.73	4-Phenoxybenzophenone

Notes: Apparatus 2, GC column 4, GC conditions 3.

4.12. Analysis of Polyacetals

Polyacetals form a different subclass of compounds with oxygen in the backbone chain. In this group are included polymers that contain the group –O–C(R_2)–O– and can be formed from the polymerization of aldehydes or ketones [30]. A typical example of a polymer from this class is paraformaldehyde or polyoxymethylene (POM). POM was discovered by Hermann Staudinger, a German chemist who received the 1953 Nobel Prize in chemistry. POM is characterized by its high strength, hardness, and rigidity to −40°C. The polymer is intrinsically opaque white, due to its high crystalline composition, but it is available in all colors. Typical applications for injection-molded POM include high-performance engineering components such as small gear wheels, eyeglass frames, ball bearings, ski bindings, fasteners, guns, knife handles, and lock systems. The material is widely used in the automotive and consumer electronics industries [121]. POM is also used for the production of consumer items like expresso coffee brewing machines and children's toys.

Figure 4.58 shows the pyrolysis–GC/MS chromatogram of POM at 700°C. The main thermal degradation product of POM is formaldehyde. Figure 4.59 shows the mass spectrum of formaldehyde.

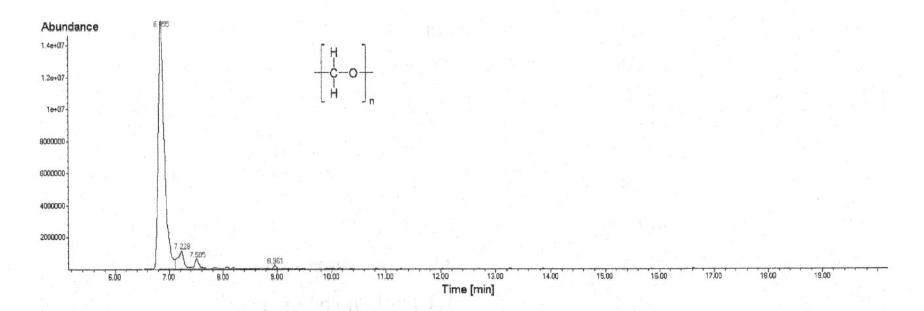

Fig. 4.58. Pyrolysis–GC/MS chromatogram of POM at 700°C. Apparatus 1, GC column 1, GC conditions 1. Peak identification: t_R = 6.85 min — formaldehyde. Mass range m/z 25–450 u.

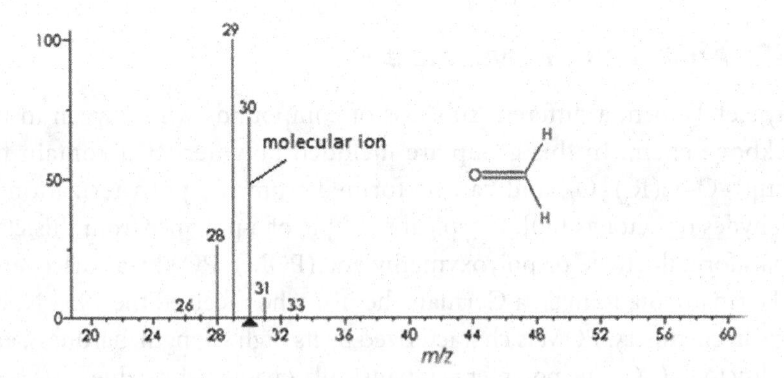

Fig. 4.59. Mass spectrum of formaldehyde identified as the main pyrolysis product of POM.

4.13. Analysis of Formaldehyde Resins

Formaldehyde resins can refer to phenol–formaldehyde resin (PF), resorcinol–formaldehyde resin (RF), melamine–formaldehyde resin (MF) and urea–formaldehyde resin (UF).

4.13.1. Characterization of PF resin

Phenolic resins are polycondensation products of phenols and aldehydes, in particular phenol and formaldehyde. The first synthetic resins and plastics were produced by Leo Baekeland in 1909 by polycondensation of phenol with formaldehyde. The first commercial plant for the production of resins was built near Berlin (Germany) in 1910 [1]. There are two broad classes of phenolic resins: novolacs and resols. Novolacs are also known as two-stage resins. They are thermoplastic polymers with an MW range of 500–3,000 g/mol [122]. Novolacs are solid at room temperature and can be melted and cooled repeatedly. They use the phenol and formaldehyde reactants in a ratio that consumes all formaldehyde leaving a chemically stable, distillable phenolic resin [122]. Novolacs are two-stage resins because an additional source of formaldehyde, or other reactant, is required to convert novolacs to useful thermoset products. Resols, on the other hand, are one-stage resins that require only the use of heat to cure the reaction mass into a crosslinked polymer. They are made by using an excess of formaldehyde under conditions that favour the formation of methylol groups or hydroxymethylation of the aromatic ring [122].

Phenolic resins are used in areas including the foundry industry, refractories, the glass and mineral wool industries, and in the manufacture of textile mats and friction linings [123]. Phenolic foam resins are used in insulation, mining, and for floral arrangements. Phenolic resins are produced for weather-proof chipboards, fiberboards as well as for decorative laminate used in indoor and outdoor applications. They are manufactured for the wooden materials industry [123].

Figure 4.60 shows the typical Py-GC/MS pyrogram obtained by pyrolysis of PF resin *Bakelite®* at 700°C with the excellent chromatographic separation of the characteristic pyrolysis products (phenol and cresol/xylenol isomers). The pyrolysis products of PF resin *Bakelite®* at 700°C are summarized in Table 4.31. Figure 4.61 shows the schematic of the main

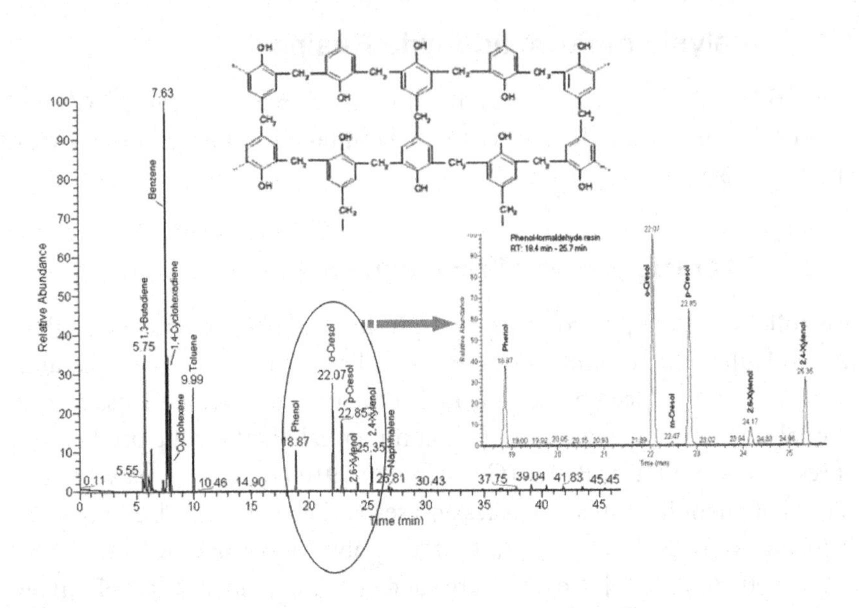

Fig. 4.60. Pyrolysis-GC/MS chromatogram of PF resin *Bakelite*® at 700°C. Apparatus 2, GC column 3, GC conditions 3, constant helium pressure of 70 kPa. For peak identification, see Table 4.31.

Table 4.31. Pyrolysis products of PF resin *Bakelite*® at 700°C.

Retention time t_R (min)	Pyrolysis product
5.75	1,3-Butadiene
7.63	Benzene
7.77	1,4-Cyclohexadiene
7.96	Cyclohexene
9.99	Toluene
18.87	Phenol
22.07	*o*-Cresol
22.85	*p*-Cresol
24.17	2,6-Xylenol
25.35	2,4-Xylenol
26.81	Naphthalene

Notes: Apparatus 2, GC column 3, GC conditions 3, constant helium pressure of 70 kPa.

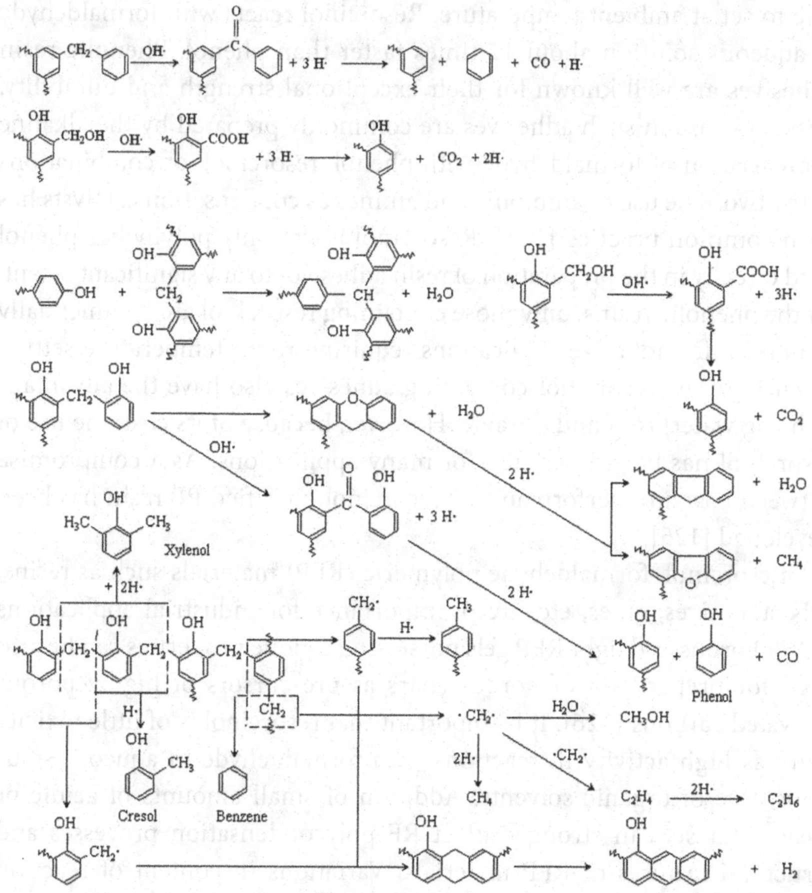

Fig. 4.61. Schematic of the main pyrolysis route of PF resin proposed by Jiang *et al.* [124]. Figure reprinted from Ref. [124] with permission from Elsevier.

pyrolysis route of PF resin, proposed by Jiang *et al.*, based on the analyses of the residual char and evolved volatiles [124].

4.13.2. Characterization of RF resin

Resorcinol-formaldehyde (RF) resin is one of the most used thermosetting adhesives in the production of exterior-grade wood structural materials. Due to the presence of the very reactive resorcinol moiety, RF resin is able to set at ambient temperature. Resorcinol reacts with formaldehyde in aqueous solution about 12 times faster than phenol. Phenolic resin adhesives are well known for their exceptional strength and durability. Resins for use in such adhesives are commonly prepared by the alkaline condensation of formaldehyde with phenol, resorcinol, or combinations of the two. The use of ammonia and amines as condensation catalysts has been common practice [125]. Resorcinol is the only polyhydric phenol used directly in the preparation of resin adhesives to any significant extent. Of the phenolic resins, only those containing resorcinol are commercially important for adhesive applications requiring room temperature setting or curing. The resorcinol-containing adhesives also have the advantage of being waterproof and durable. However, because of its cost, the use of resorcinol has been restricted for many applications. As a compromise between cost and performance, a resorcinol modified PF resin has been developed [125].

Resorcinol–formaldehyde polymeric (RFP) materials such as resins, gels, adhesives, glues, etc. are of importance for industrial applications [126]. Porous and light RFP gels possessing unique properties can be also used for preparation of porous chars as pre-cursors of highly porous activated carbons [126]. It is important that resorcinol is of little toxicity and has high activity in reactions with formaldehyde in aqueous solutions, free of organic solvents. Addition of small amounts of acidic or basic catalysts can strongly affect RF polycondensation processes and structural features of RFP materials. Variations in content of reaction components and water as a solvent, and changes in a catalyst type and content, allow significant and controlled changes in the structural, morphological, and textural characteristics of the RFP gels. Frequently, carbonates or hydroxides of sodium or potassium are used as catalysts to prepare RFP materials [126].

RF resin therapy, commonly known as "Russian red" cement, has been a unique method of endodontic therapy in Eastern Europe, Russia, China, and other countries around the world. RF resin is a combination of formaldehyde/alcohol, resorcinol powder, and a sodium hydroxide catalyst. It is mixed to various consistencies and placed in root canals as a temporary or permanent obturating material. The methods for using RF therapy were described in 1957 and have been widely used since 1960 [127].

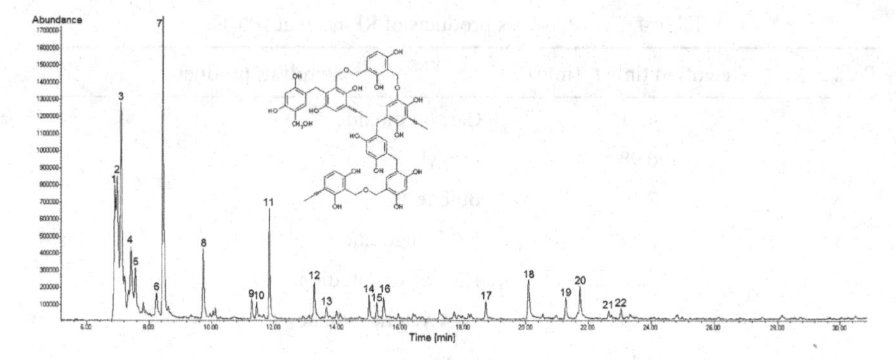

Fig. 4.62. Pyrolysis–GC/MS chromatogram of RF resin at 700°C. Apparatus 1, GC column 2, GC conditions 1. For peak identification, see Table 4.32.

Figure 4.62 shows the pyrolysis–GC/MS chromatogram of RF resin at 700°C. The degradation products of RF resin at 700°C are summarized in Table 4.32. Figure 4.63 shows the mass spectrum of resorcinol (1,3-benzenediol).

4.13.3. Characterization of MF resin

Melamine-formaldehyde or melamine resin (MF) is a hard, very durable, and versatile thermosetting plastic (aminoplast) with good fire and heat resistance. It is made from melamine and formaldehyde by condensation of the two monomers. Its good fire-retardant properties are due to the release of nitrogen gas when burned or charred. MF resins are primarily made up of melamine and formaldehyde with formaldehyde acting as the crosslinker. Melamine reacts with formaldehyde under slightly alkaline conditions to form mixtures of various methylolmelamines [128].

Melamine resins are used for the manufacture of many products, including kitchenware, laminates, overlay materials, particleboards, and floor tiles. Melamine is also used in the manufacture of flame-resistant materials. These include textiles such as upholstery, firemen uniforms, thermal liners, and heat-resistant gloves and aprons [128]. Melamine and its salts are also used as fire-retardant additives in paints, plastics, and paper. Melamine foam is a special form of melamine resin. It finds uses as an insulating and soundproofing material and more recently as a cleaning

Table 4.32. Pyrolysis products of RF resin at 700°C.

Peak No.	Retention time t_R (min)	Pyrolysis product
1	6.91	Carbon dioxide
2	6.98	Propylene
3	7.10	Butene
4	7.42	1,4-Pentadiene
5	7.56	1,3-Cyclopentadiene
6	8.24	1,3-Cyclohexadiene
7	8.45	Benzene
8	9.74	Toluene
9	11.29	Ethylbenzene
10	11.44	*p*-Xylene
11	11.86	Styrene
12	13.29	Phenol
13	13.67	α-Methylstyrene
14	15.02	*o*-Cresol
15	15.26	Indene
16	15.47	*p*-Cresol
17	18.72	Naphthalene
18	20.09	1,3-Benzenediol (Resorcinol)
19	21.28	2-Methyl-1,3-benzenediol (2-Methylresorcinol)
20	21.73	5-Methylresorcinol (Orcinol)
21	22.64	4,5-Dimethylresorcinol
22	23.05	Biphenyl

Notes: Apparatus 1, GC column 2, GC conditions 1.

abrasive. The foam products can be used for removing scuffs and dirt from a wide range of surfaces. Furthermore, some filters are made without of porous melamine. These filters can be used in hot environments and are extremely efficient [128].

Figure 4.64 shows the pyrolysis–GC/MS chromatogram of MF resin at 700°C. The pyrolysis products of MF resin at 700°C, identified by using the *NIST 05* mass spectra library are summarized in Table 4.33.

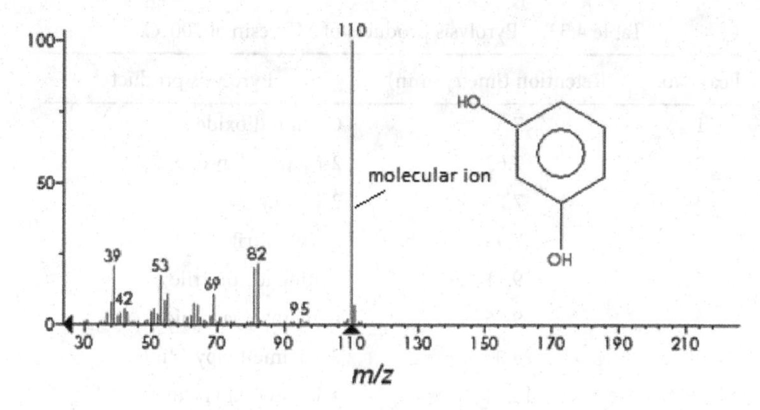

Fig. 4.63. Mass spectrum of resorcinol (1,3-benzenediol).

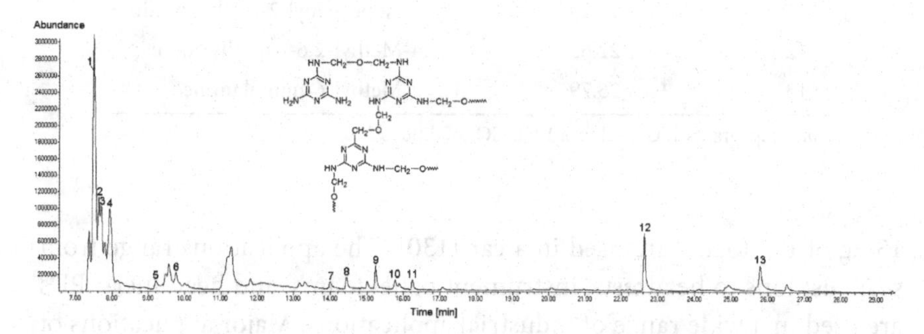

Fig. 4.64. Pyrolysis–GC/MS chromatogram of MF resin at 700°C. Apparatus 1, GC column 1 UI, GC conditions 2. For peak identification, see Table 4.33.

4.14. Analysis of Polyurethanes (PU)

In 1935, Otto Bayer (I.G. Farben, Leverkusen, Germany) discovered that diisocyanates undergo a rapid reaction with glycols to produce PU in a polyaddition reaction [1]. PU are formed by chemical reaction between a di/poly isocyanate and a diol or polyol, forming repeating urethane groups, generally, in presence of a chain extender, catalyst, and/or other additives. Often, ester, ether, urea, and aromatic rings are also present, along with urethane linkages in PU backbone [129]. PUs are a large family of polymers whose composition has evolved over time. PU foams are largely used in automobiles, home furniture, and thermal insulation. Approximately

Table 4.33. Pyrolysis products of MF resin at 700°C.

Peak No.	Retention time t_R (min)	Pyrolysis product
1	7.50	Carbon dioxide
2	7.63	2-Methyl-1-propene
3	7.70	2-Butene
4	7.93	Acrylonitrile
5	9.21	Aminoacetonitrile
6	9.76	Dimethylcyanamide
7	14.00	2,4-Dimethylpyridine
8	14.44	4-Isopropylpyridine
9	15.24	5-Ethenyl-2-methylpyridine
10	15.78	2,3-Cyclopentenopyridine
11	16.25	5-Isopropenyl-2-methylpyridine
12	22.63	4-Methyl-2,6-naphthyridine
13	25.79	2-Methyl-6-quinolamine

Notes: Apparatus 1, GC column 1 UI, GC conditions 2.

13 kg of PU foams are used in a car [130]. The applications range from seat cushions to headrests, instrument panel foams, and headliners. PUs are used in a wide range of industrial applications. Major applications of flexible PU slab foams are in furniture, carpet underlay and bedding, and as foam in transportation. They are the most useful reaction polymers for the footwear industry. Microcellular PU elastomers are used in the production of shoe soles for every type of footwear. Over the years, PUs became the materials of choice for applications like all-weather athletic areas, outdoor games areas, children's playgrounds, tennis courts, and multi-sports halls. PUs are also widely present in museum collections such as natural history museums, fine art museums, modern art museums, either as a part of the artifacts, or as a material for their conservation (stuffing, protection, packaging, and storage) [129]. Among museum collections, they are featured predominantly as sculptures, design objects, cushioning materials, textiles, and toys.

Pyrolysis mechanisms and characterization of PU polymers have been studied by many authors [96, 130, 131]. The results demonstrate that thermal degradation of such polymeric systems generates a wide range of products like gases (e.g. CO, CO_2, NH_3, and HCN), aliphatic and aromatic hydrocarbons, carbonyl compounds, amines, and phenyl isocyanates/diisocyanates [96, 131]. Figure 4.65 shows the thermal degradation mechanisms of PUs proposed by Hiltz [131]. The pyrolysis–GC/MS chromatogram of a steering wheel from the automotive industry, at 700°C, identified as PU synthesized by the polyaddition reaction of diphenylmethane-4,4′-diisocyanate (MDI) and poly(propylene glycol) is shown in Fig. 4.66.

Fig. 4.65. Thermal degradation mechanisms of PUs proposed by Hiltz [131]. Reprinted from Ref. [131] with permission from Elsevier.

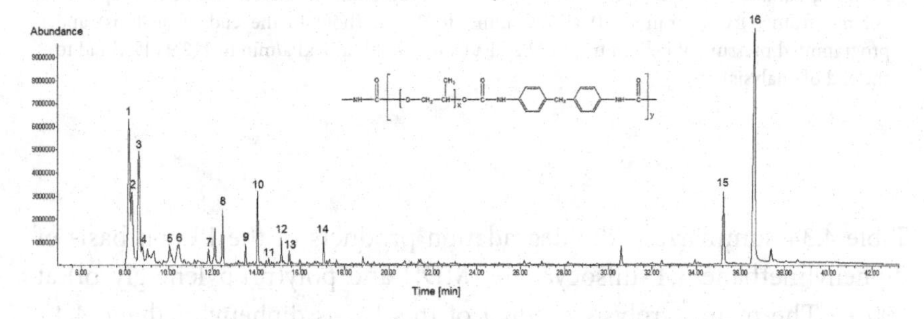

Fig. 4.66. Pyrolysis–GC/MS chromatogram of a steering wheel PU on basis of diphenylmethane-4,4′-diisocyanate (MDI) and poly(propylene glycol) at 700°C. Apparatus 1, GC column 1, GC conditions: programmed temperature of the capillary column from 60°C (1-min hold) at 7°C/min to 280°C (hold to the end of analysis) and programmed pressure of helium from 122.2 kPa (1-min hold) at 7 kPa/min to 212.9 kPa (hold to the end of analysis). For peak identification, see Table 4.34.

Table 4.34. Pyrolysis products of a steering wheel PU on basis of diphenylmethane-4,4′-diisocyanate (MDI) and poly(propylene glycol) at 700°C.

Peak No.	Retention time t_R (min)	Pyrolysis product
1	8.16	Propylene
2	8.30	Propane + acetaldehyde
3	8.62	Acetone
4	8.78	Acrylonitrile
5	10.04	1-Hydroxy-2-propanone
6	10.46	Ethylene glycol
7	11.82	Toluene
8	12.46	1-(1-Methylethoxy)-2-propanone
9	13.49	Ethylbenzene
10	14.03	Styrene
11	14.55	Cumene
12	15.13	Isocyanatobenzene (Carbanil)
13	15.50	α-Methylstyrene
14	17.05	*p*-Tolylisocyanate
15	35.23	Diphenylmethane-2,4′-diisocyanate or Diphenylmethane-2,2′-diisocyanate
16	36.64	Diphenylmethane-4,4′-diisocyanate (MDI)

Notes: Apparatus 1, GC column 1, GC conditions: programmed temperature of the capillary column from 60°C (1-min hold) at 7°C min^{-1} to 280°C (hold to the end of analysis) and programmed pressure of helium from 122.2 kPa (1-min hold) at 7 kPa/min to 212.9 kPa (hold to the end of analysis).

Table 4.34 summarized the degradation products of the PU on basis of diphenylmethane-4,4′-diisocyanate (MDI) and poly(propylene glycol) at 700°C. The main pyrolysis product of this PU is diphenylmethane-4,4′-diisocyanate (Fig. 4.66, peak 16, Table 4.34).

Figure 4.67 shows the pyrolysis–GC/MS chromatogram at 700°C of PU synthesized from 2,4-toluene diisocyanate (TDI) and 1,4-butanediol. The pyrolysis products identified by using the *NIST 05* mass spectra library are summarized in Table 4.35.

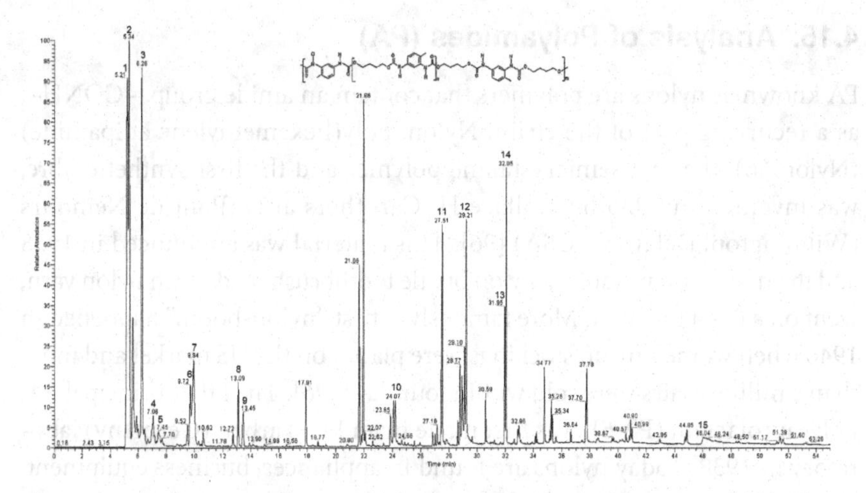

Fig. 4.67. Pyrolysis–GC/MS chromatogram at 700°C of PU synthesized from 2,4-toluene diisocyanate (TDI) and 1,4-butanediol. Apparatus 2, GC column 3, GC conditions 3. For peak identification, see Table 4.35.

Table 4.35. Pyrolysis products of PU synthesized from 2,4-toluene diisocyanate (TDI) and 1,4-butanediol. Pyrolysis temperature 700°C.

Peak No.	Retention time t_R (min)	Pyrolysis product
1	5.21	Propylene
2	5.34	Butene
3	6.26	Butanal
4	7.06	1-Butanol
5	7.45	4-Ethoxy-1-butene
6	9.72	Butanedial
7	9.94	2-Hydroxytetrahydrofuran
8	13.09	4-Butoxy-1-butene
9	13.45	1,1′-Oxybisbutane
10	24.07	4-Butoxy-1-butanol
11	27.51	2-Methyltetrahydro-2-furanol
12	29.21	2,4-Toluene diisocyanate (TDI)
13	31.95	Butyric anhydride
14	32.05	1-Ethoxy-4-(vinyloxybutoxy)butane
15	46.04	4,4′-Methylene-bis(2-chloroaniline) (crosslinking agent)

Notes: Apparatus 2, GC column 3, GC conditions 3.

4.15. Analysis of Polyamides (PA)

PA known as nylons are polymers that contain an amide group, -CONH–, as a recurring part of the chain. Nylon, poly(hexamethylene adipamide) (Nylon 66), the first semicrystalline polymer and the first synthetic fibre, was invented in 1935 by Wallace H. Carothers at DuPont de Nemours (Wilmington, Delaware, USA) [96]. This material was announced in 1938 and the first nylon product, a nylon bristle toothbrush made with nylon yarn, went on sale in this year. More famously, a first "nylon-boom" happened in 1940 when women nylon stockings were placed on the US market and more than a million pairs were sold within four days [96]. The other principal PA, polycaprolactam (PA 6), was first made in an I.G. Farben (Germany) laboratory in 1938. Today nylons are found in appliances, business equipment, electrical/electronic devices, furniture, hardware, machinery, packaging, and transportation [96]. Transportation is the largest market for nylons. The softer nylons are used in fuel lines, air brake hoses, and coatings. Industrial applications are attracted to the excellent fatigue resistance and repeated impact strength boots, ice and roller skate supports, bicycle wheels, kitchen utensils, garden equipment, toys, and fishing lines [96].

Figure 4.68 shows the pyrolysis–GC/MS chromatogram obtained after pyrolysis of polycaprolactam (PA 6) at 550°C. The pyrolysis products of PA 6 at 550°C are summarized in Table 4.36. As shown in Fig. 4.68 and Table 4.36, the most abundant compound resulting from the thermal

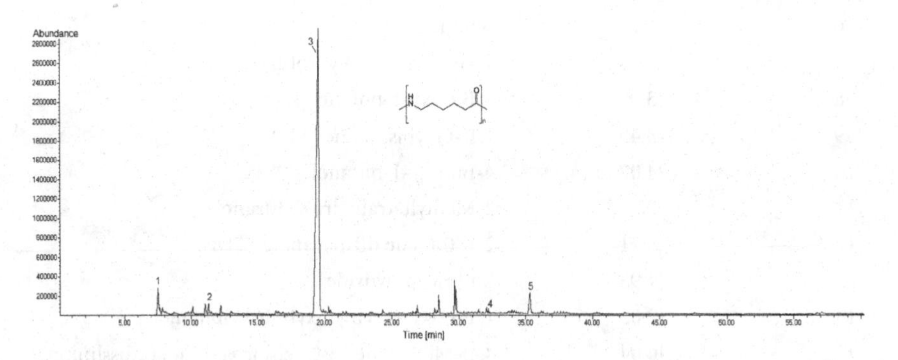

Fig. 4.68. Pyrolysis–GC/MS chromatogram of polycaprolactam (PA 6) at 550°C. Apparatus 1, GC column 1 UI, GC conditions 2. For peak identification, see Table 4.36.

Table 4.36. Pyrolysis products of polycaprolactam (PA 6) at 550°C.

Peak No.	Retention time t_R (min)	Pyrolysis product
1	7.51	Carbon dioxide
2	11.34	Hexanenitrile
3	19.54	ε-Caprolactam (monomer)
4	32.22	Hexadecanenitrile
5	35.33	1,8-Diazacyclotetradecane-2,9-dione (dimer)

Notes: Apparatus 1, GC column 1 UI, GC conditions 2.

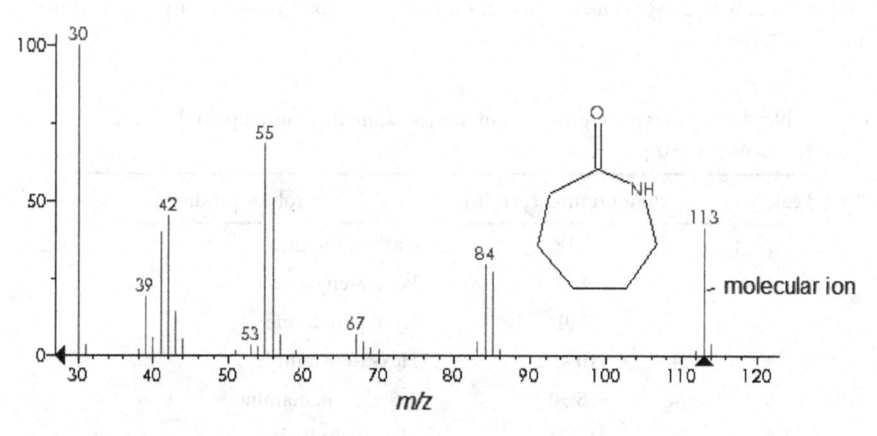

Fig. 4.69. Mass spectrum of ε-caprolactam identified as the main pyrolysis product of polycaprolatam (PA 6).

degradation of PA 6 is ε-caprolactam (monomer and dimer). Figure 4.69 shows the mass spectrum of ε-caprolactam, identified as the main pyrolysis product of polycaprolatam.

Figure 4.70 shows the pyrolysis–GC/MS chromatogram obtained after pyrolysis of poly(hexamethylene adipamide) (PA 66, Nylon 66) at 550°C. The pyrolysis products of PA 66 at 550°C are summarized in Table 4.37. As shown in Fig. 4.70 and Table 4.37, the most abundant compound resulting from the thermal degradation of PA 66 is cyclopentanone, which is characteristic of adipic acid-based PA. Figure 4.71 shows the mass spectrum of cyclopentanone identified as the main pyrolysis product of poly(hexamethylene adipamide).

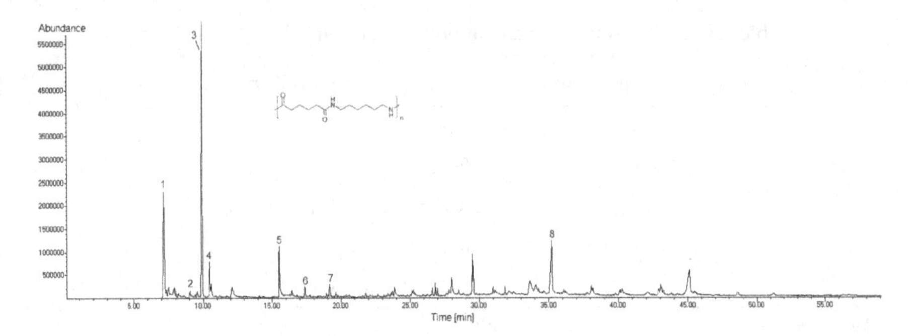

Fig. 4.70. Pyrolysis–GC/MS chromatogram of poly(hexamethylene adipamide) (PA 66, Nylon 66) at 550°C. Apparatus 1, GC column 1 UI, GC conditions 2. For peak identification, see Table 4.37.

Table 4.37. Pyrolysis products of poly(hexamethylene adipamide) (PA 66, Nylon 66) at 550°C.

Peak No.	Retention time t_R (min)	Pyrolysis product
1	7.18	Carbon dioxide
2	9.05	Pent-4-enylamine
3	9.91	Cyclopentanone
4	10.43	Hex-5-enylamine
5	15.50	1,6-Hexanediamine
6	17.34	Hexanedinitrile
7	19.16	Caprolactam
8	35.23	1,8-Diazacyclotetradecane-2,7-dione

Notes: Apparatus 1, GC column 1 UI, GC conditions 2.

4.16. Analysis of Polyaramid

Polyaramids are generally prepared by the reaction between an aromatic amine group and an aromatic acid–halide group. Simple AB homopolymers may be prepared like in Eq. (4.2) [132]:

$$n\,\mathrm{NH_2{-}Ar{-}COCl} \rightarrow \mathrm{{-}(NH{-}Ar{-}CO)_n{-}} + n\,\mathrm{HCl} \qquad (4.2)$$

The most well-known aramids (*Kevlar*, *Twaron*, *Nomex*, *New Star* and *Teijinconex*) are AABB polymers. *Nomex*, *Teijinconex*, and *New Star*

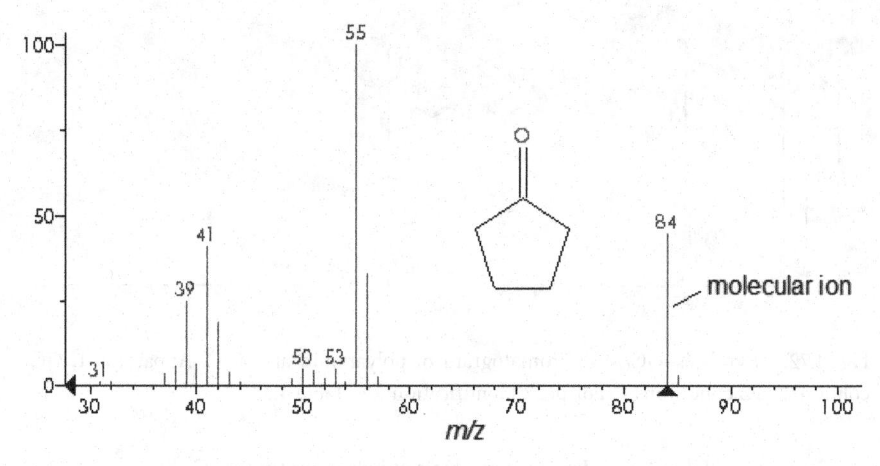

Fig. 4.71. Mass spectrum of cyclopentanone identified as the main pyrolysis product of poly(hexamethylene adipamide) (PA 66).

contain predominantly the meta-linkage and are poly(*m*-phenylene isophthalamide)s (MPIA). *Kevlar* and *Twaron* are both poly(*p*-phenylene terephthalamide)s (PPTA), the simplest form of the AABB *para*-polyaramide. PPTA is a product of *p*-phenylene diamine (PPD) and terephthaloyl dichloride (TDC or TCL).

Polyaramid fibers are a class of heat-resistant and strong synthetic fibers. They are used in aerospace and military applications, for ballistic rated, body armor, fabric and ballistic composites, in bicycle tires, and as an asbestos substitute. Figure 4.72 shows the pyrolysis–GC/MS chromatogram of the polyaramid (PPTA) fibers pyrolyzed at 900°C. The decomposition products are summarized in Table 4.38. The identified main degradation products of polyaramid at 900°C are benzene (peak 3, t_R = 7.56), aniline (peak 7, t_R = 11.61 min), and benzonitrile (peak 8, t_R = 11.81 min) [133].

4.17. Analysis of Polyphenylene Sulfide [Poly(thio-*p*-phenylene)]

Polyphenylene sulfide (PPS) is an organic polymer discovered by Charles Friedel and James Mason Crafts in 1888 [134]. The polymer is consisting of aromatic rings linked with sulfides (see the chemical structure in Fig. 4.73).

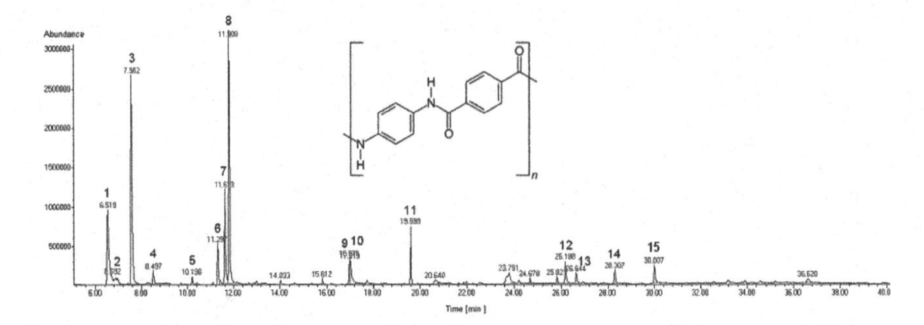

Fig. 4.72. Pyrolysis–GC/MS chromatogram of polyaramid at 900°C. Apparatus 1, GC column 2, GC conditions 1. For peak identification, see Table 4.38.

Table 4.38. Pyrolysis products of polyaramid at 900°C.

Peak No.	Retention time t_R (min)	Pyrolysis product
1	6.52	Carbon dioxide
2	6.70	Acrylonitrile
3	7.56	Benzene
4	8.50	Toluene
5	10.20	Styrene
6	11.30	Isocyanatobenzene
7	11.61	Aniline
8	11.81	Benzonitrile
9	16.98	1,2-Benzodinitrile
10	17.02	Benzene-1,4-diamine
11	19.60	Biphenyl
12	26.19	Acridine
13	26.64	1,1′-Biphenyl-4-amine
14	28.31	Carbazole
15	30.01	*N*-Phenylbenzamide

Notes: Apparatus 1, GC column 2, GC conditions 1.

PPS is a versatile material that gives extruded and molded components the ability to meet exceptionally demanding criteria. This semicrystalline engineering thermoplastic has outstanding thermal stability, superior toughness, inherent flame resistance, and excellent chemical resistance [134]. It also

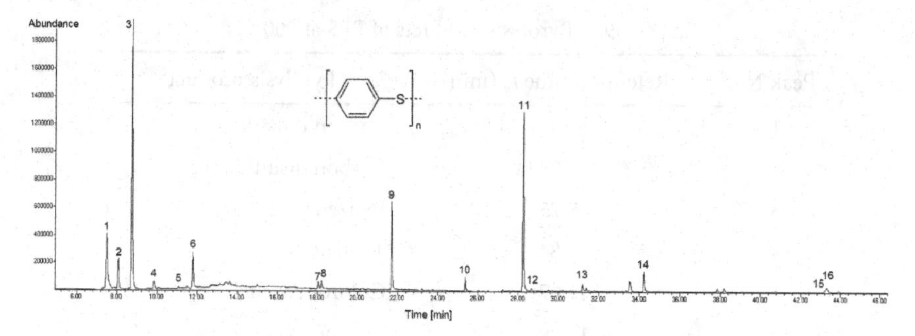

Fig. 4.73. Pyrolysis–GC/MS chromatogram of PPS at 900°C. Apparatus 1, GC column 1 UI, GC conditions 2. For peak identification, see Table 4.39.

has high mechanical strength, impact resistance, and dimensional stability as well as good electrical properties. Regular PPS is a linear polymeric material of modest MW (18,000 g/mol). PPS possesses high temperature resistance combined with good mechanical properties, exceptional chemical and solvent resistance, high-dimensional stability, and easy processing. A commercially important preparation procedure of PPS involves the polycondensation of 1,4-dichlorobenzene and sodium sulfide in polar solvent, such as *N*-methylpyrrolidone, at high temperature and pressure [134]. PPS is primarily used in injection molding, precision mechanical parts, electrical components, and components for use in harsh environments. Manufactured parts from PPS are engine components, valves, high-pressure nozzles, fuel manifolds exterior light reflectors, coil bobbins, and flow measuring units for hot and cold water [134]. Regular PPS is primarily used in coatings due to its thermal and chemical stability. Cured PPS gives good results in coatings and injection-molding compounds, whereas linear PPS is ideal for fibers and tougher injection-molding compounds due to the increased molecular chain length resulting in higher tenacity, elongation, and impact strength. Due to some improved mechanical properties, as well as tenacity and ductility, branched PPS is suitable for films and injection-molding compounds [134].

Figure 4.73 shows the pyrolysis–GC/MS chromatogram of PPS at 900°C. Table 4.39 shows the degradation products of PPS at 900°C identified by using the *NIST 05* mass spectra library. Figures 4.74 and 4.75 show the mass spectra of diphenyl sulfide and dibenzothiophene, respectively. Both compounds are characteristic pyrolysis products of PPS.

Table 4.39. Pyrolysis products of PPS at 900°C.

Peak No.	Retention time t_R (min)	Pyrolysis product
1	7.51	Carbon dioxide
2	8.08	Carbon disulfide (CS_2)
3	8.75	Benzene
4	9.84	Toluene
5	11.07	Chlorobenzene
6	11.80	Styrene
7	18.05	Naphthalene
8	18.21	2-Benzothiophene
9	21.70	Biphenyl
10	25.37	Diphenyl sulfide
11	28.26	Dibenzothiophene
12	28.67	9-Methylene-9H-fluorene
13	31.20	2-Phenylnaphthalene
14	34.22	*p*-Terphenyl
15	43.20	1,4-Bis(phenylthio)benzene
16	43.30	4-Phenylbenzothiophene

Notes: Apparatus 1, GC column 1 UI, GC conditions 2.

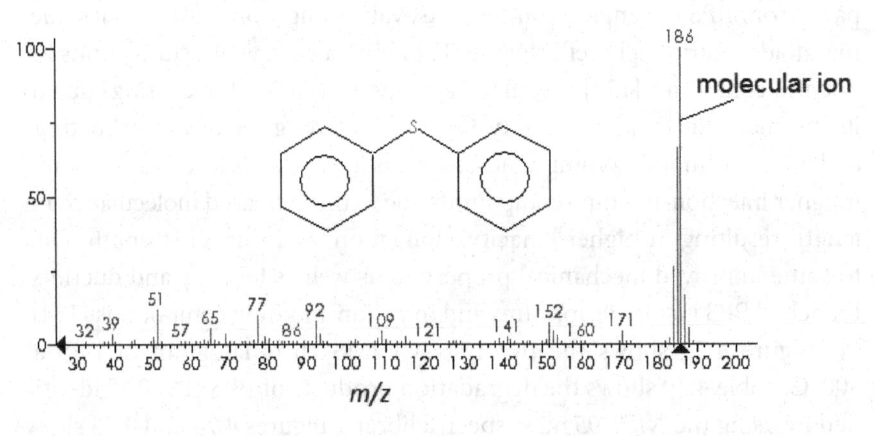

Fig. 4.74. Mass spectrum of diphenyl sulfide.

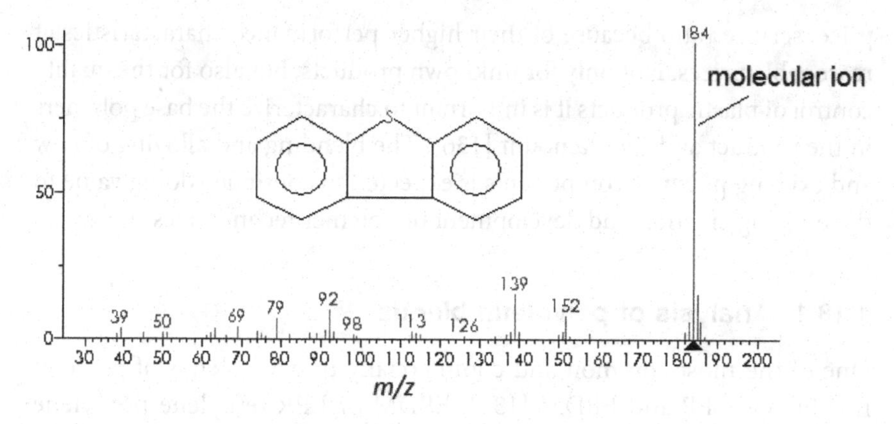

Fig. 4.75. Mass spectrum of dibenzothiophene.

4.18. Pyrolysis–GC/MS of Polymer Blends

Polymers are ideal materials to meet developing societal needs due to their broad performance profiles, economical shape forming ability and generally low density [135]. They thus can provide valuable properties in a cost-effective manner. The blending and alloying of various types of polymers, along with the compounding in of other ingredients such as fillers, plasticizers, stabilizers, colorants, and reinforcements, extend their reach by providing a means to economically generate polymer systems with broader performance spectra than possible from individual polymer components [135]. While some commercial blends are marketed for wide-ranging applications deriving from their generally beneficial mechanical, electrical, thermal, chemical, or flow properties, as the technology matures, it focuses more on new blends and alloys, having unique or critical properties or property mixes for specific applications [135]. In order to reach specific chemical and physical properties of a plastic product, different compositions of polymer blends are produced by the combination and variation of the ratio of two or more homopolymers and copolymers and/or the usage of additional components like additives and plasticizers. Since polymer blends offer many useful properties due to the combination of properties of the individual components, these types of products are playing an important role in materials development and application. They provide an optimal price–service ratio because of their higher performance characteristics at reasonable prices. Not only for unknown products, but also for the quality control of plastic products it is important to characterize the base polymers in the product and their amount [136]. The blending and alloying of new and existing polymer components is expected to continue adding value to the ongoing creation and development of polymer technologies.

4.18.1. Analysis of polyolefin blends

One of the most common and commercially utilized blends of polyolefins involves PP and EPDM [137]. EPDM or EPR (ethylene-propylene rubber) has been added to PP as an impact modifier for over a decade. HDPE can be added to the PP/EPR blend to achieve maximum toughness. Semi-flexible PP/EPR blends (60–80 wt.% EPR) were commercialized by Uniroyal, DuPont, and B. F. Goodrich in the mid-1970s [137]. Applications cited for these blends were wire and cable insulation, automotive bumpers and fascia, hose, gaskets, seals, and weather stripping. These blends replaced plasticized PVC and crosslinked rubbers where improved low temperature flexibility, rubbery properties, and thermoplastic character were desired. Similar commercial applications were also noted for blends of HDPE and butyl rubber [137]. The addition of ethylene copolymers (ethylene/vinyl acetate, ethylene/ethyl acrylate), EPR or butyl rubber to either LDPE or HDPE has been commercially utilized to improve environmental stress crack resistance, toughness, filler acceptance, film tear resistance, improved flexibility, etc. Many of these blends are proprietary formulations and have not been specifically disclosed by the manufacturers. Polyolefin blend technology is of critical importance to the success of material suppliers to the various application areas served by these materials [137]. Even conventional polymer blends can find new or expanded applications under evolving codes and performance requirements. For example, white roofing membranes of PP/EPR thermoplastic olefin elastomer having high reflectance result in reduced cooling costs, while blending PS for foam insulation with an olefinic polymer can enhance its elastic strength and toughness against fracture during installation [135].

Figure 4.76 shows the pyrolysis–GC/MS chromatogram of a PE/PP blend at 700°C as well as the identification of the pyrolysis products.

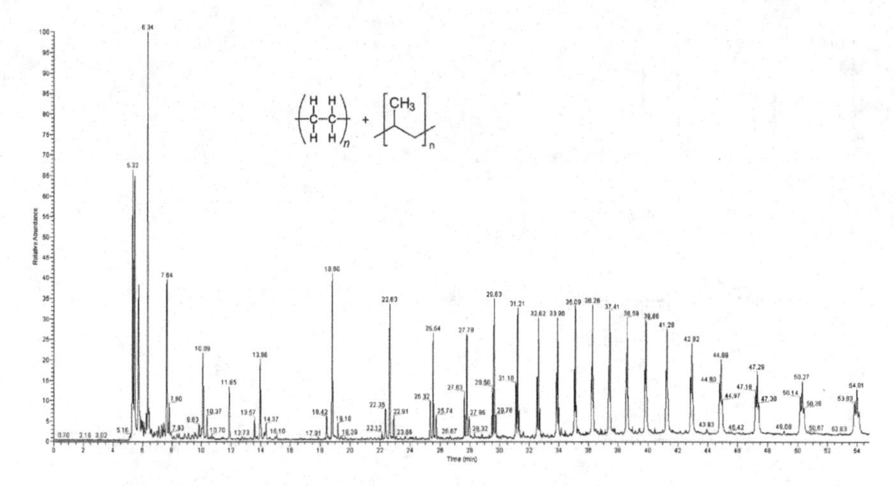

Fig. 4.76. Pyrolysis–GC/MS chromatogram of a PE/PP blend at 700°C. Apparatus 2, GC column 3, GC conditions 3. Peak identification: t_R = 5.32 min — propylene, t_R = 6.29 min — 2-methyl-1-pentene (propylene dimer), t_R = 11.85 min — 2,4-dimethyl-1-heptene (propylene trimer) — all characteristic for the pyrolysis of PP. Other peaks consist of serial triplets, corresponding to C_4–C_{28} α,ω-alkadienes, α-alkenes, and *n*-alkanes, respectively, in the order of increasing *n*+1 carbon number in the molecule, characteristic for the pyrolysis of PE.

4.18.2. Analysis of PP/PS blends

A key feature of the PP/PS alloys is the impact strength/stiffness envelope that reportedly exceeds the performance of conventional PP and approach in some aspects with those of other engineering resins such as polyacetals, PC/ABS or PC/PBT [138]. Compounding varying levels of glass fibers and reinforcing mineral filers provides the desired balance of stiffness and toughness in these PP/PS alloys. The reinforced grades reportedly exhibit improved stiffness and creep resistance compared to PP alone and may compete against reinforced PA and polyesters, particularly in applications that do not require high temperature performance. Typical applications in development with these PP/PS alloys include automotive bumper beams, pillars, sporting and recreational equipment, sledge hammer handles, and other consumer tools and appliance components [138].

Figure 4.77 shows the pyrolysis–GC/MS chromatogram of a PP/PS blend and the comparison with the pyrograms of PP and PS at 700°C.

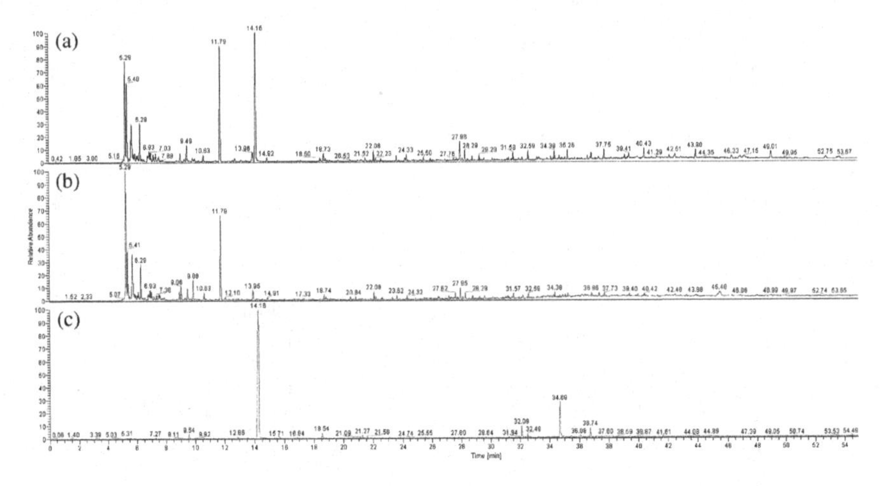

Fig. 4.77. Pyrolysis–GC/MS chromatogram of (a) PP/PS blend, (b) PP, (c) PS at 700°C. Apparatus 2, GC column 3, GC conditions 3. Peak identification: t_R = 5.29 min — propylene, t_R = 6.29 min — 2-methyl-1-pentene (propylene dimer), t_R = 11.79 min — 2,4-dimethyl-1-heptene (propylene trimer), t_R = 14.16 min — styrene.

4.18.3. Analysis of ABS blends

4.18.3.1. *Characterization of ABS/PC blends*

Polymer blends are used among others in healthcare technology. Drug delivery is a growing area of the healthcare technology. Needleless injection techniques and inhalation are specific modes of delivery that are fueling this growth, driven largely by elevated requirements for patient comfort. As with blood handling applications, biocompatibility and sterilizability are key requirements. However, impact and wear resistance are also important. Some of the blends used in these applications include PC/PBT and ABS/PC. Sleep therapy and respiratory care represent an additional class of applications in healthcare. Respirators, ventilators, and positive-pressure devices, to allow airways to function properly, are specific examples and require biocompatibility. Respiratory masks and valves require chemical resistance and impact performance. PC-based blends are commonly used in applications in this space [135].

One of the more significant segments in the healthcare industry involves the handling and management of fluids. This segment includes blood handling during surgery, general blood collection, and blood oxygenation. The segment also includes membrane and filter applications such as leukocyte filters, arterial filters, and kidney dialysis. In addition to blood handling, intravenous (IV) and gastrointestinal fluid delivery systems often involve bottles, bags, pumps, and tubes. Blood handling typically requires clarity and transparency, but applications such as IV fluid handling are more applicable to blend systems that have impact resistance and chemical resistance. Polyester blends and ABS/PC blends have seen penetration into some of these applications [135].

The display industry has evolved considerably over recent years. ABS/PC blends impart lightweight and impact/chemical resistance along with good processing for molded cases incorporating portable electronics such as cell phones. Soft plastics can also be molded against and bonded to the hard blend for gripping and aesthetics.

ABS/PC blends targets interior automotive applications, for which it is optimally formulated to provide sufficient heat resistance along with low-temperature ductility to meet safety requirements.

Figure 4.78 shows the pyrolysis–GC/MS chromatogram of ABS and PC blend (ABS/PC) at 700°C. The identified pyrolysis products of the ABS/PC blend are summarized in Table 4.40.

4.18.3.2. *Characterization of ABS/PA 6 blends*

PA are polymers of interest in engineering issues because of their ideal strength and stiffness, low friction, and chemical and wear resistance [139].

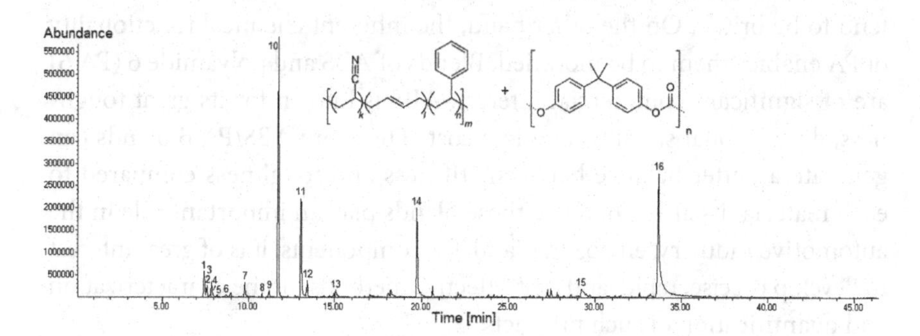

Fig. 4.78. Pyrolysis–GC/MS chromatogram of ABS and PC blend (ABS/PC) at 700°C. Apparatus 1, GC column 1 UI, GC conditions 2. For peak identification, see Table 4.40.

Table 4.40. Pyrolysis products of ABS and PC blend (ABS/PC) at 700°C.

Peak No.	Retention time t_R (min)	Pyrolysis product
1	7.46	Carbon dioxide
2	7.61	1,3-Butadiene
3	7.78	Acetonitrile
4	7.89	Acrylonitrile (2-Propenenitrile)
5	8.21	Methacrylonitrile (2-Methyl-2-propenenitrile)
6	8.71	Benzene
7	9.80	Toluene
8	10.80	4-Ethenylcyclohexene
9	11.22	Ethylbenzene
10	11.78	Styrene
11	13.10	Phenol
12	13.46	α-Methylstyrene
13	15.09	4-Methylphenol (*p*-Cresol)
14	19.73	*p*-Isopropenylphenol
15	29.24	4-(1-Methyl-1-phenylethyl)-phenol (4-Cumylphenol)
16	33.70	4,4′-(1-Methylethylidene)bisphenol (Bisphenol A)

Notes: Apparatus 1, GC column 1 UI, GC conditions 2.

However, at low temperatures and under extreme loading conditions they tend to be brittle. On the other hand, the inherent chemical functionality of PA enables them to be modified. Blends of ABS and polyamide 6 (PA 6) are of significant commercial interest. ABS is known for its great toughness, dimensional stability, and low cost. Therefore ABS/PA 6 blends can generate a better balance between stiffness and toughness compared to each material itself [139]. Since these blends play an important role in the automotive industry, enclosures, and TV components, it is of great interest to develop precise, rapid, and cost-effective methods for the characterization and quantification of such products.

Figure 4.79 shows the pyrolysis–GC/MS chromatogram of a ABS and PA 6 blend at 700°C. Table 4.41 shows the identified, by using the *NIST 05*

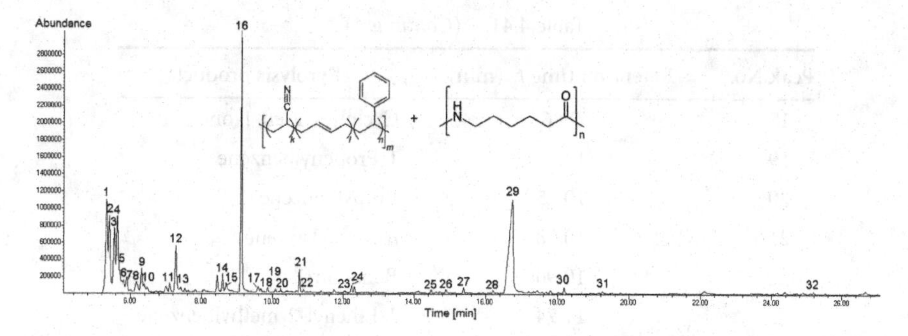

Fig. 4.79. Pyrolysis–GC/MS chromatogram of ABS (20% w/w) and PA 6 (80% w/w) blend at 700°C. Apparatus 1, GC column 1, GC conditions 1. For peak identification, see Table 4.41.

Table 4.41. Pyrolysis products of ABS (20% w/w) and PA 6 (80 % w/w) blend at 700°C.

Peak No.	Retention time t_R (min)	Pyrolysis product
1	5.33	Propene
2	5.41	1,3-Butadiene
3	5.55	Acetonitrile
4	5.63	Acrylonitrile
5	5.70	1,3-Cyclopentadiene
6	5.81	Propanenitrile
7	5.91	Methacrylonitrile
8	6.20	2-Butenenitrile
9	6.32	Benzene
10	6.48	2-Pentenenitrile
11	7.10	3-Methyl-3-butenenitrile
12	7.28	Toluene
13	7.54	Cyclopentanone
14	8.59	Ethylbenzene
15	8.70	Hexanenitrile
16	9.12	Styrene
17	9.67	1-Ethyl-4-methylbenzene

(*Continued*)

Table 4.41. (*Continued*)

Peak No.	Retention time t_R (min)	Pyrolysis product
18	9.86	Cyclohex-2-en-1-one
19	10.09	1-Propenylbenzene
20	10.25	Propylbenzene
21	10.78	α-Methylstyrene
22	10.90	Benzonitrile
23	11.74	1-Ethenyl-2-methylbenzene
24	12.25	1H-Indene
25	14.48	Methyl-1H-indene isomer
26	14.90	Methyl-1H-indene isomer
27	15.39	Naphthalene
28	16.21	3-Phenylpropanenitrile
29	16.75	Caprolactam (Azepan-2-one)
30	18.18	4-Phenylbutanenitrile
31	19.26	1,1′-Biphenyl
32	25.19	Stilbene

Notes: Apparatus 1, GC column 1, GC conditions 1.

mass spectra library degradation products of a ABS and PA 6 blend at 700°C. Figure 4.80 shows the mass spectra of the compounds characteristic for the pyrolysis of ABS/PA 6 blends.

In the early work of the author (P. K.) and his coworkers [136], blends of ABS and PA 6 were characterized and quantified by Py–GC/MS and Py–GC/FID at 700°C, respectively. Figure 4.81 shows the Py–GC/MS chromatograms (fivefold determination) of the ABS/PA 6 sample at 700°C The obtained analytical data of the peak area ratios were evaluated and led to the conclusion that the Py–GC/FID method is more suitable for quantification of the investigated samples [136]. This method has been validated successfully, evaluating parameters such as linearity and working range, accuracy, limits of detection (LOD) and quantification (LOQ), and

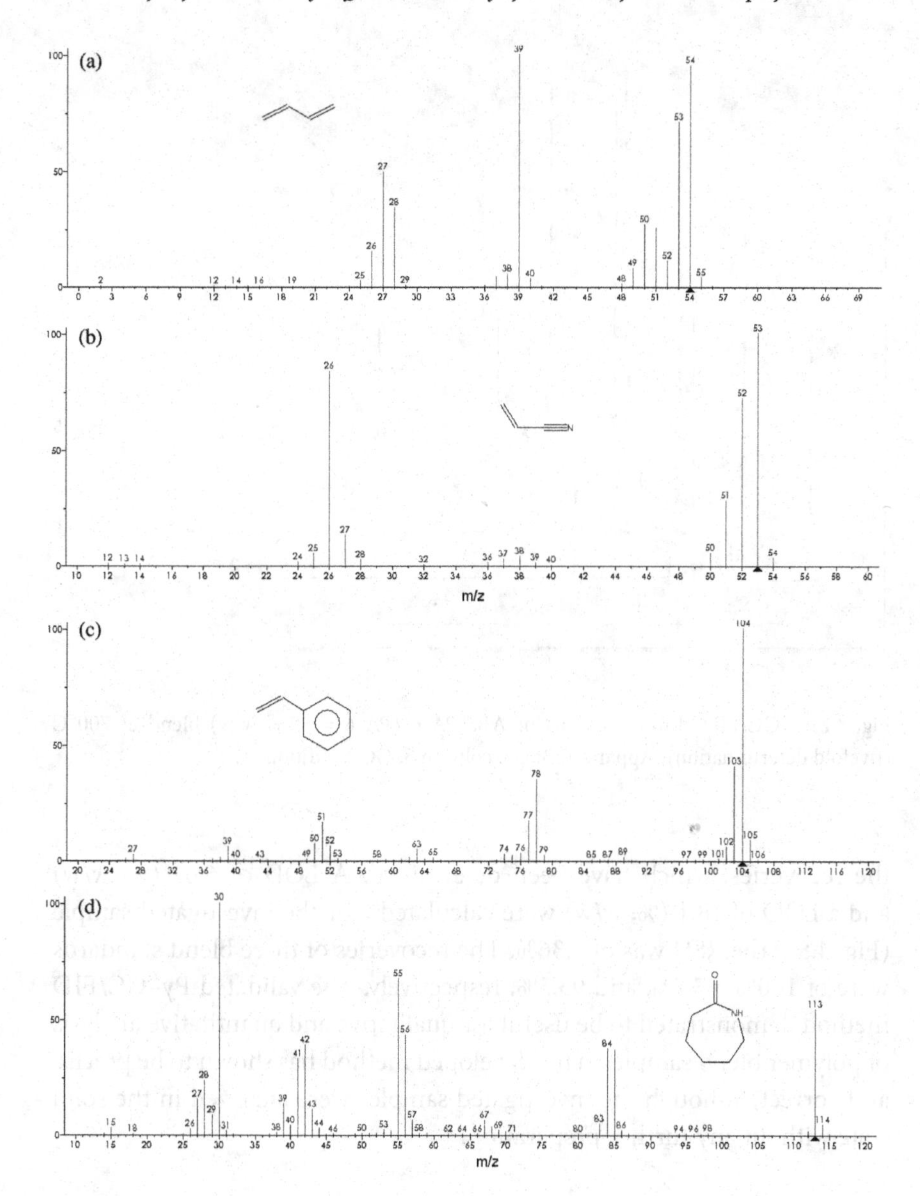

Fig. 4.80. Mass spectra of the significant pyrolysis products of the ABS/PA 6 blend at 700°C: (a) 1,3-butadiene, (b) acrylonitrile, (c) styrene, (d) caprolactam.

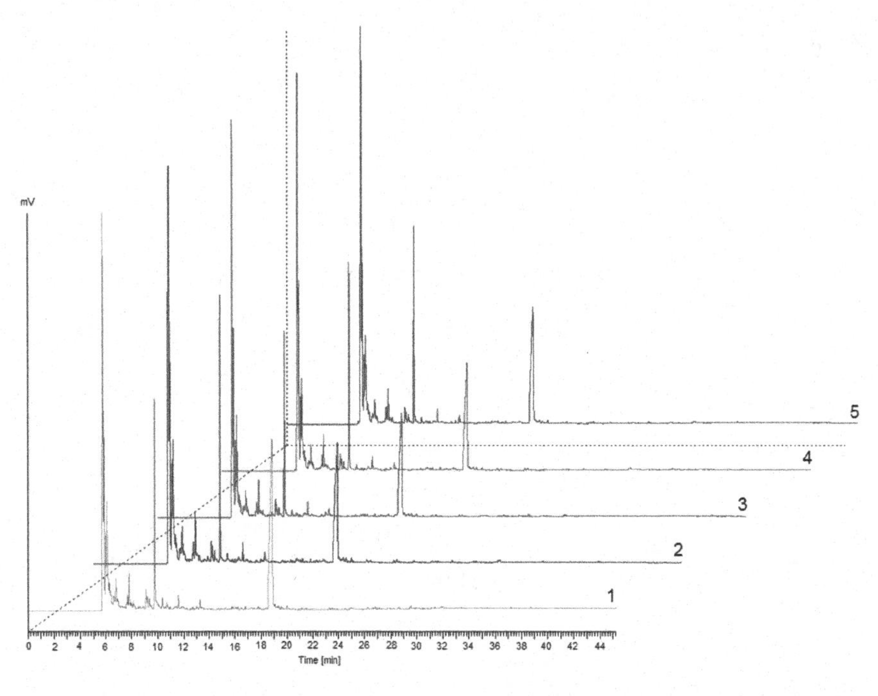

Fig. 4.81. GC/FID chromatograms of ABS/PA 6 (PA 6 = 60%, w/w) blend at 700°C (fivefold determination). Apparatus 3, GC column 5, GC conditions 5.

the recoveries, which have been determined. A LOD of 6.02 (%, w/w) and a LOQ of 18.1 (%, w/w) were calculated. For the investigated sample (Fig. 4.81) the RSD was of 2.36%. The recoveries of three blend standards were of 100%, 97.7%, and 93.3%, respectively. The validated Py–GC/FID method demonstrated to be useful for qualitative and quantitative analysis of polymer blend samples. This developed method has shown to be precise and correct, although the investigated samples were analyzed in the solid state without any further preparation.

4.18.4. Analysis of PVC and SMMA blend

Polymer blends are of considerable importance, as blending could provide a means for improving the impact strength of fragile polymers. PVC belongs to a major class of engineering plastics, that possess many unique properties that are suitable for a wide variety of technical and industrial products. It is available at a relatively low cost, is nonflammable, and has good chemical resistance and corrosion resistance. However, the stiff and brittle material is deficient for PVC expanded applications. PVC is blended with rubbery polymers to improve its impact strength, thus making it suitable for rigid application. It is well known that graft copolymers formed during the production of rubber-modified plastics serve to promote adhesion between the rubbery phase and glassy phase, together contributing to the high impact strength of the blend. Furthermore, the addition of block or graft copolymers in relatively small quantities further improves mechanical properties [140].

Figure 4.82 shows the pyrolysis–GC/MS chromatogram of PVC and SMMA [poly(styrene-*co*-methyl methacrylate)] blend at 700°C. The pyrolysis products of this blend and the retention data of the compounds identified by using the *NIST 05* mass spectra library are summarized in Table 4.42.

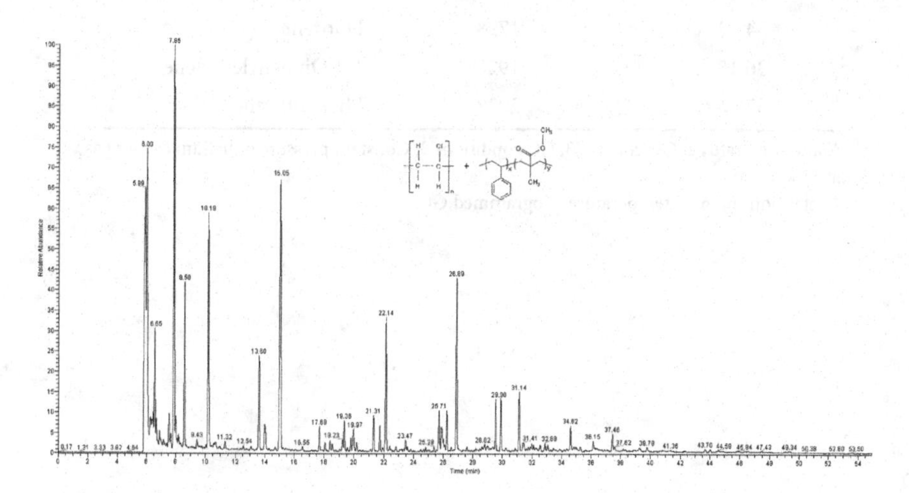

Fig. 4.82. Pyrolysis–GC/MS chromatogram of PVC and SMMA blend at 700°C. Apparatus 2, GC column 3, GC conditions 3. Constant pressure of helium carrier gas at 70 kPa. For peak identification, see Table 4.42.

Table 4.42. Pyrolysis products of PVC and SMMA blend at 700°C.

Retention time t_R (min)	Retention index*	Pyrolysis product
5.89	235	Hydrogen chloride + propene
6.00	253	2-Butene
6.55	340	1,3-Cyclopentadiene
7.85	547	Benzene
8.58	668	Methyl methacrylate (MMA)
10.19	744	Toluene
13.60	839	Ethylbenzene
14.04	849	*p*-Xylene
15.05	874	Styrene
21.31	975	1-Propenylbenzene
22.14	988	Indene
25.71	1213	1-Methylindene
26.89	1276	Naphthalene
29.54	1423	2-Methylnaphthalene
29.90	1447	1-Methylnaphthalene
31.14	1533	1,1′-Biphenyl
34.62	1798	Fluorene
36.15	1926	1,2-Diphenylethylene
37.46	2028	Phenanthrene

Notes: Apparatus 2, GC column 3, GC conditions 3. Constant pressure of helium carrier gas at 70 kPa.

*Retention index in temperature-programmed GC.

5

Py–GC/MS of Biopolymers

Biomass is a complex material, mainly composed of cellulose, hemicelluloses, and lignin, in addition to extractives (tannins, fatty acids, resins) and inorganic salts. This renewable raw material is of great potential. It can be used for energy generation and for production of high-value chemicals. Biomass resources that can be used for energy production cover a wide range of materials such as forestry residues, energy crops, organic wastes, agricultural residues, etc. [141, 142]. The chemical structure and major organic components in biomass are extremely important in biomass pyrolysis processes. Knowledge of the pyrolysis characteristics of the three main components is the basis, and thus essentially important for a better understanding of biomass thermal chemical conversion. Fast pyrolysis, which is known as a promising process to convert pretreated biomass to bio-oil, is affected by the biomass types and reaction conditions. It is an effective method to convert biomass into high-value gaseous, liquid, and solid fuels. Because this biomass fast pyrolysis is an endothermic process and therefore requires an external energy input, considerable attention has been paid to reducing the energy consumption during the pyrolysis process [143].

5.1. Analysis of Lignin

Lignin, nature's dominant aromatic polymer, is found in most terrestrial plants in the approximate range from 15% to 40% dry weight and provides structural integrity. It is the most recalcitrant of the three components of lignocellulosic biomass (cellulose, hemicellulose and lignin) and accounts usually for 10–25 wt.% of biomass. It is predominantly obtained from cooking liquors produced in pulping processes [143]. The advent of new cellulosic bio-refineries, however, will introduce an excess supply of different, nonsulfonated, native, and transgenically modified lignin into the process streams. Therefore, the efficient utilization of lignin resources is necessary for the paper-making industry and in cellulosic bio-refinery. Besides cellulose and hemicellulose, lignin is the most abundant organic natural raw material and expected to play a key role regarding the world-wide production of bio-based products. It is part of the cell walls of lignocellulosic plants, responsible for maintaining their rigidity and resistance against environmental conditions [144]. Pulp and paper industries produce large quantities of lignin (ca. 55×10^6 tons per year) as a side product of delignification via pulping processes [144]. The two most important industrial paper technologies are Kraft and sulphite pulping, leading to sulfur-containing degraded lignin fractions which are predominantly used as a secondary energy source. The utilization of lignin for the synthesis of high-value added materials is an issue of economic and environmental importance. There are studies of the utilization of lignin in phenolic-formaldehyde resins which show comparable properties to the commercially produced resins [144].

Lignin is a complex phenylpropanoid polymer that derives mainly from the oxidative condensation of three *p*-hydroxycinnamyl alcohol monomers differing in their degree of methoxylation: *p*-coumaryl, coniferyl, and sinapyl alcohols. These monolignols produce the *p*-hydroxyphenyl (H), guaiacyl (G), and syringyl (S) units when incorporated into the lignin polymer [145]. The resulting structure of lignin formed by dehydrogenative polymerization of these monolignols is an extremely complex 3D macromolecule. Lignin is thermally stable. Its decomposition starts at 145°C and is the most intensive in the range from 325 to 535°C. Figure 5.1 shows the thermal decomposition reaction of lignin proposed by Sobiesiak *et al.* [146].

Figure 5.2 shows the pyrolysis–GC/MS chromatogram of Organosolv isolated lignin from beech wood at 550°C. The results of identification are summarized in Table 5.1. Fragments derived from the syringyl (S), guaicacyl (G), and *p*-hydroxyphenyl (H) units are also reveals in Table 5.1.

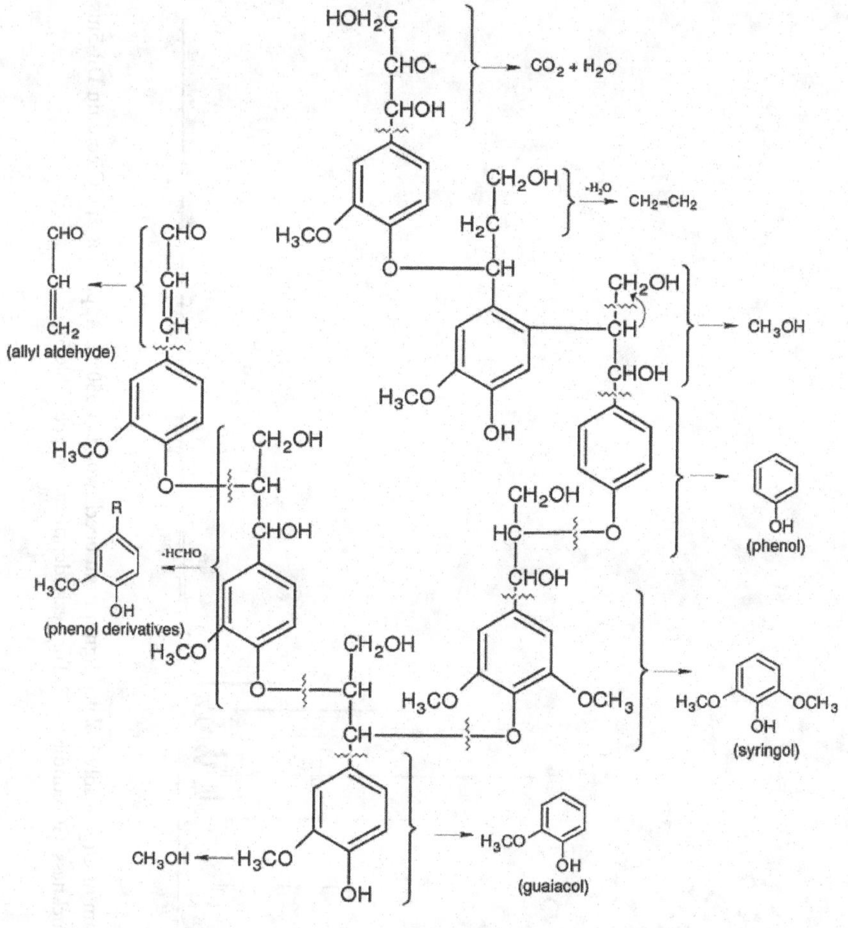

Fig. 5.1. Thermal decomposition of lignin. Figure reprinted from Ref. [146] with permission from Elsevier.

5.2. Analysis of Cotton Fibers

Cotton fibers are nature's purest form of cellulose, composing about 90% of α-cellulose [147]. Cellulose is the nature's most abundant polymer. The noncellulosics are located on the outer layers or inside the lumens of the fibers, whereas the secondary cell wall is purely cellulose. Both, the structure and composition of the cellulose and noncellulosics depend on the variety and the growing conditions [147]. The pyrolysis products

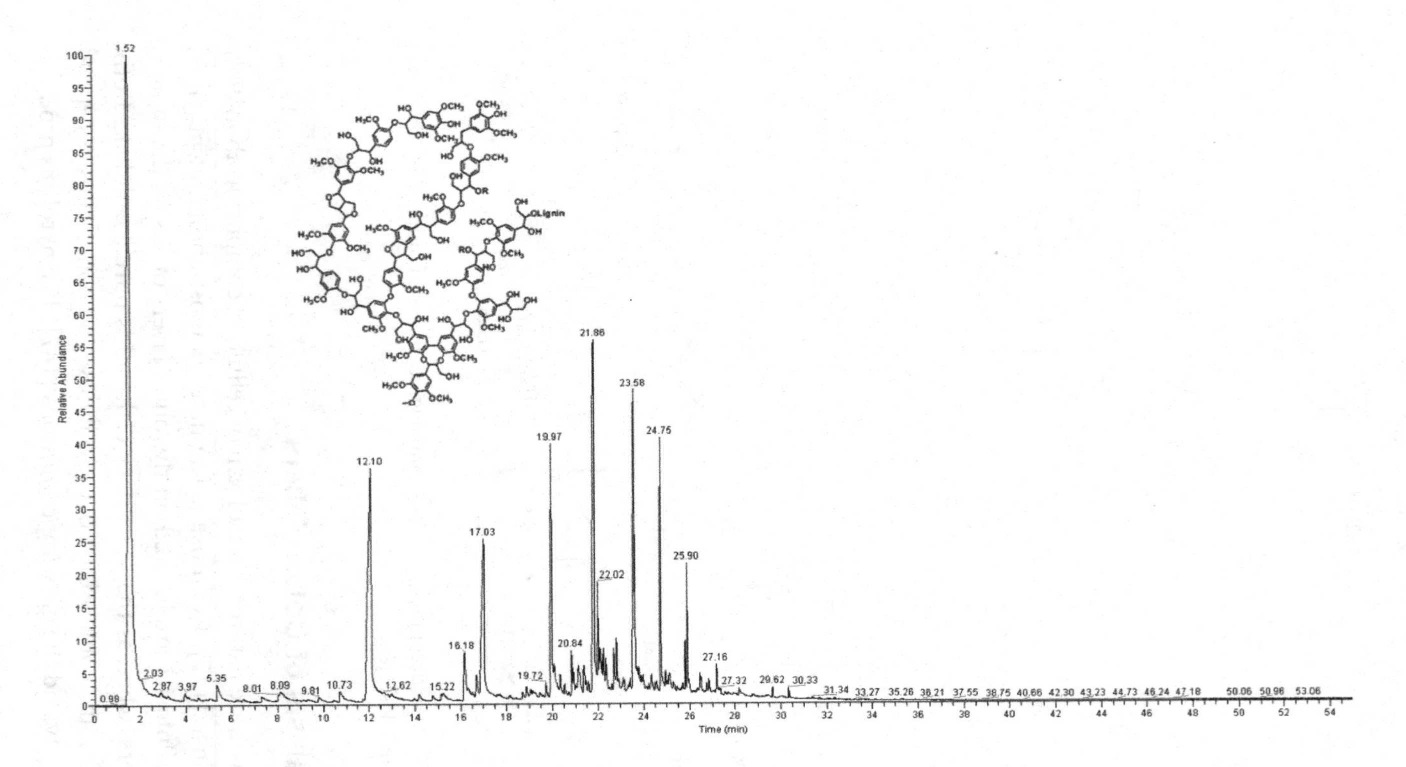

Fig. 5.2. Pyrolysis–GC/MS chromatogram of Organosolv isolated lignin from beech wood at 550°C. Apparatus 2, GC column DB-5ms, 30 m long, 0.25 mm I. D., 0.25 μm film thickness, GC conditions 3. For peak identification, see Table 5.1.

Table 5.1. Compounds identified by pyrolysis–GC/MS of Organosolv isolated lignin from beech wood at 550°C.

Compound	Retention time t_R (min)	Molecular weight (g mol^{-1})	Origin
Carbon dioxide	1.52	44	
Phenol	12.10	94	L-H
2-Methylphenol (*o*-cresol)	16.18	108	L–H
4-Methylphenol (*p*-cresol)	17.03	108	L–H
2,3-Dimethylphenol	21.41	122	L–G
4-Ethylphenol	21.86	122	L–G
4-Ethyl-2-methylphenol	22.02	136	L–G
4-Ethyl-3-methylphenol	22.27	136	L–G
2-Methoxy-4-vinylphenol	22.81	150	L–G
Vanillin	23.58	152	L–G
2-Methoxy-4-propylphenol (4-propylguaiacol)	24.75	166	L–G
Trimethoxybenzene	25.40	168	L–S
2-Methoxy-4-(1-propenyl)phenol (isoeugenol)	25.90	164	L–G
3′,5′-Dimethoxyacetophenon	26.43	180	L–G
2,6-Dimethoxy-4-(2-propenyl)phenol (4-allylsyringol)	27.16	194	L–S
3,5-Dimethoxy-4-hydroxycinnamaldehyde (sinapinaldehyde)	29.62	208	L–S

Notes: Apparatus 2, GC column DB-5ms, 30 m long, 0.25 mm I. D., 0.25 μm film thickness, GC conditions 3.

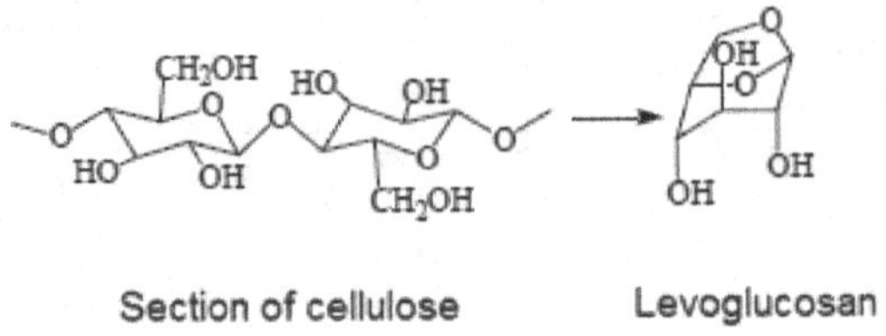

Section of cellulose Levoglucosan

Fig. 5.3. Chemical reaction of the pyrolytic unzipping of cellulose.

of cellulose at 550–600°C, identified by GC/MS, include a number of substituted phenols, ketonic compounds, acids, methyl esters, furans, pyrans, anhydrosugars, and hydrocarbons. The major pyrolysis products are levoglucosan (1,6-anhydro-ß-D-glucopyranose) and substituted furans [33]. Levoglucosan is known to be an important intermediate in the pyrolysis of cellulose (see Fig. 5.3) [148]. Figure 5.4 shows the obtained from our laboratory pyrolysis–GC/MS chromatogram of cotton fibers at 700°C. As can be seen from Fig. 5.4, the main pyrolysis product of cotton fibers at 700°C is carbon dioxide (CO_2). The retention data of the pyrolysis products of cotton fibers at 700°C are summarized in Table 5.2.

5.3. Analysis of Wood of Tropic Trees

The need to shift from a fossil-based society toward a renewable and sustainable one has necessitated the thermal degradation of carbon-neutral biomass as a source of chemicals and fuels. Pyrolysis of two tropical woods, namely *Brastegia eurycoma* and *Nauclea latifolia* from Southeast Nigeria was investigated. After trituration, the woods samples were subjected to direct pyrolysis–GC/MS at 550°C in helium atmosphere. Figure 5.5 shows the obtained pyrolysis–GC/MS chromatograms of the investigated wood samples at 550°C. The degradation products of *Brastegia eurycoma* and *Nauclea latifolia* woods are summarized in the Table 5.3. Most of the volatile untreated wood pyrolyzates were phenolic and aldehyde compounds. On the basis of the pyrolysis results, it can be concluded that wood can serve as an ideal lingo-chemical alternative for petrochemicals.

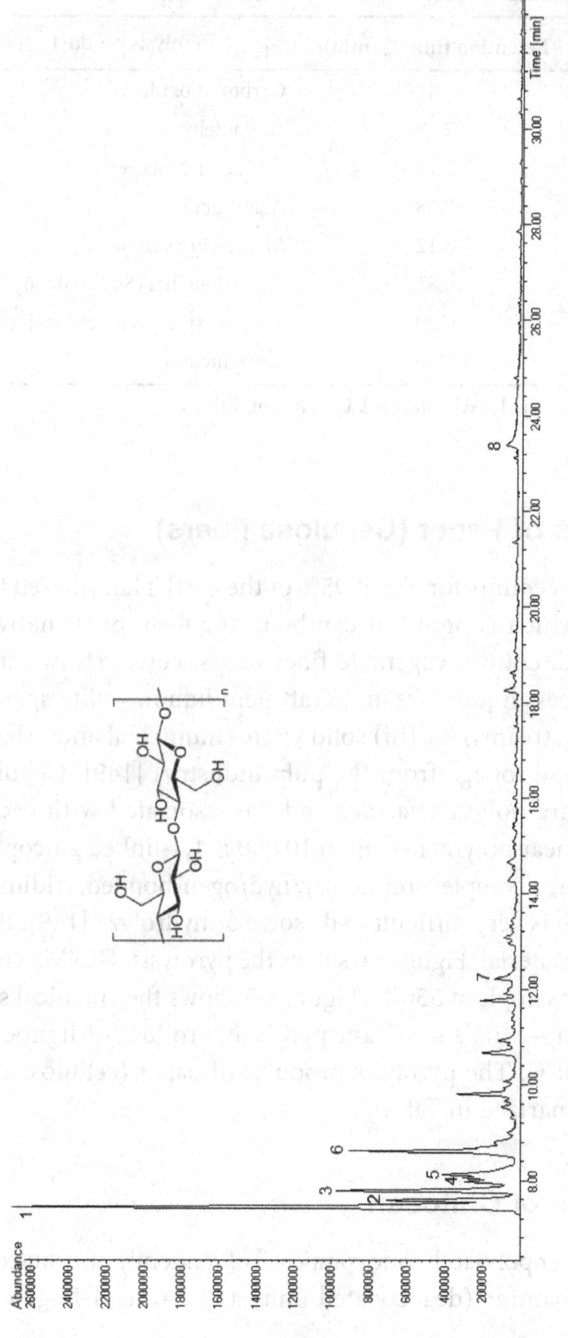

Fig. 5.4. Pyrolysis–GC/MS chromatogram of cotton fibers at 700°C. Apparatus 1, GC column 1 UI, GC conditions 2. For peak identification, see Table 5.2.

Table 5.2. Pyrolysis products of cotton fibers at 700°C.

Peak No.	Retention time t_R (min)	Pyrolysis product
1	7.44	Carbon dioxide
2	7.58	Acetaldehyde
3	7.78	1-Propen-2-ol acetate
4	7.98	Acetic acid
5	8.12	Methylvinylketone
6	9.81	1,4-Butanedial (Succinaldehyde)
7	12.29	2-Hydroxy-2-cyclopenten-1-one
8	23.39	Levoglucosan

Notes: Apparatus 1, GC column 1 UI, GC conditions 2.

5.4. Analysis of Paper (Cellulose fibers)

Lignocellulose accounts for about 95% of the earth's land-based biomass, about 25% of which is lignin. It can be in the form of (i) native lignocellulose (wood, cotton, vegetable fiber crops, cereal straws, and corn stalks), (ii) process liquor stream (Kraft pent liquor, sulfite spent liquor, pulp mill waste streams), or (iii) solid waste (municipal and urban, cattle manure, bark and foliage from the pulp industry) [149]. Cellulose and hemicellulose are polysaccharides and are associated with each other. Cellulose is a linear polymer of up to 10,000 ß-1,4-linked glucopyranosyl units arranged in a complex, refractory, hydrogen-bonded, tridimensional structure, which is very difficult to dissolve or hydrolyze [149]. Paper is a lignocellulosic material. Figure 5.6 shows the pyrolysis–GC/MS chromatogram of a paper sample at 550°C. Figure 5.7 shows the chemical structure of levoglucosan — the significant pyrolysis product of lignocellulosic materials at 550°C. The pyrolysis products of paper (cellulose fibers) at 550°C are summarized in Table 5.4.

5.5. Analysis of Chitosan

Chitosan is a linear polysaccharide composed of randomly distributed β-(1,4)-linked D-glucosamine (deacetylated unit) and *N*-acetyl-D-glucosamine

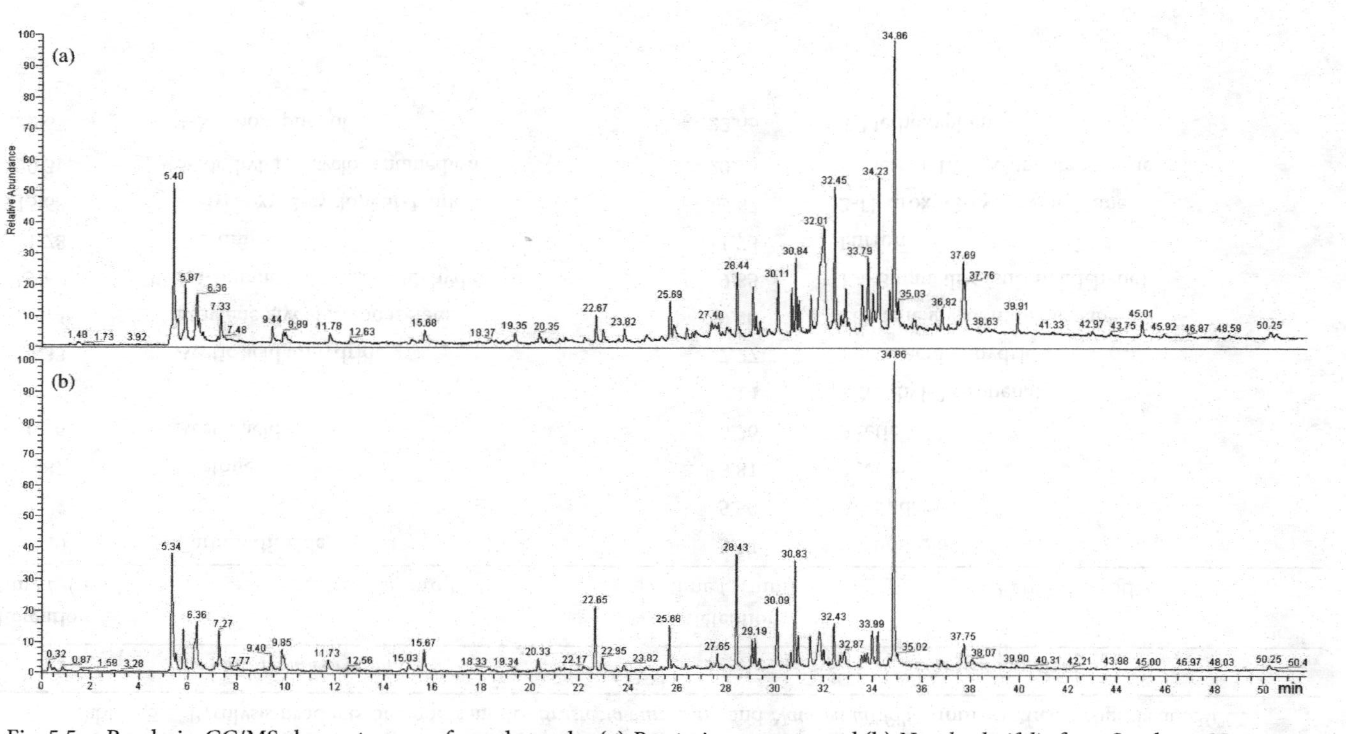

Fig. 5.5. Pyrolysis–GC/MS chromatogram of wood samples (a) *Brastegia eurycoma* and (b) *Nauclea latifolia* from Southeast Nigeria at 550°C. Apparatus 2, GC column 3, GC conditions 3. For peak identification, see Table 5.3.

Table 5.3. Pyrolysis products of wood samples *Brastegia eurycoma* and *Nauclea latifolia* from Southeast Nigeria at 550°C.

Brastegia eurycoma		*Nauclea latifolia*	
Retention time t_R (min)	Pyrolysis product	Retention time t_R (min)	Pyrolysis product
5.40	Carbon dioxide	5.34	Carbon dioxide
		5.50	Acetaldehyde
5.87	Acetone	5.81	Acetone
6.36	Acetic acid	6.36	Acetic acid
		7.14	2-Methyl-2-propenal
7.33	Acetic acid anhydride	7.27	Acetic acid anhydride
9.44	Ethylene glycol monoacetate	9.40	Ethylene glycol monoacetate
9.89	1,4-Butanedial (succinaldehyde)	9.85	1,4-Butanedial (succinaldehyde)
11.78	Furfural	11.73	Furfural
15.68	2-Hydroxy-2-cyclopent-1-one	15.67	2-Hydroxy-2-cyclopent-1-one
20.35	3-Methyl-1,2-cyclopentanedione	20.33	3-Methyl-1,2-cyclopentanedione
22.67	4-Methoxyphenol	22.65	4-Methoxyphenol

25.69	2-Methoxy-4-methylphenol	25.68	2-Methoxy-4-methylphenol
		27.65	4-Ethyl-2-methoxyphenol (*p*-ethylguaiacol)
28.44	2-Methoxy-4-vinylphenol	28.43	2-Methoxy-4-vinylphenol
29.09	2,6-Dimethoxyphenol	29.07	2,6-Dimethoxyphenol
30.11	4-Hydroxy-3-methoxybenzaldehyde (vanillin)	30.09	4-Hydroxy-3-methoxybenzaldehyde (vanillin)
30.84	2-Methoxy-4-propenylphenol	30.83	2-Methoxy-4-propenylphenol
30.97	2-Methoxy-4-propylphenol	30.95	2-Methoxy-4-propylphenol
31.46	4′-Hydroxy-3′-methoxyacetophenone	31.44	4′-Hydroxy-3′-methoxyacetophenone
31.85–32.01	Levoglucosan	31.84	Levoglucosan
32.45	2-*tert*-Butyl-4-methoxyphenol	32.43	2-*tert*-Butyl-4-methoxyphenol
33.79	4-Hydroxy-3,5-dimethoxybenzaldehyde		
34.23	2,6-Dimethoxy-4-(2-propenyl)-phenol	34.22	2,6-Dimethoxy-4-(2-propenyl)-phenol
34.67	4′-Hydroxy-3′,5′-dimethoxyacetophenone		
34.86	4-[(1E)-3-Hydroxy-1-propenyl]-2-methoxyphenol	34.86	4-[(1E)-3-Hydroxy-1-propenyl]-2-methoxyphenol
37.69	3,5-Dimethoxy-4-hydroxycinnamaldehyde	37.69	3,5-Dimethoxy-4-hydroxycinnamaldehyde

Notes: Apparatus 2, GC column 3, GC conditions 3.

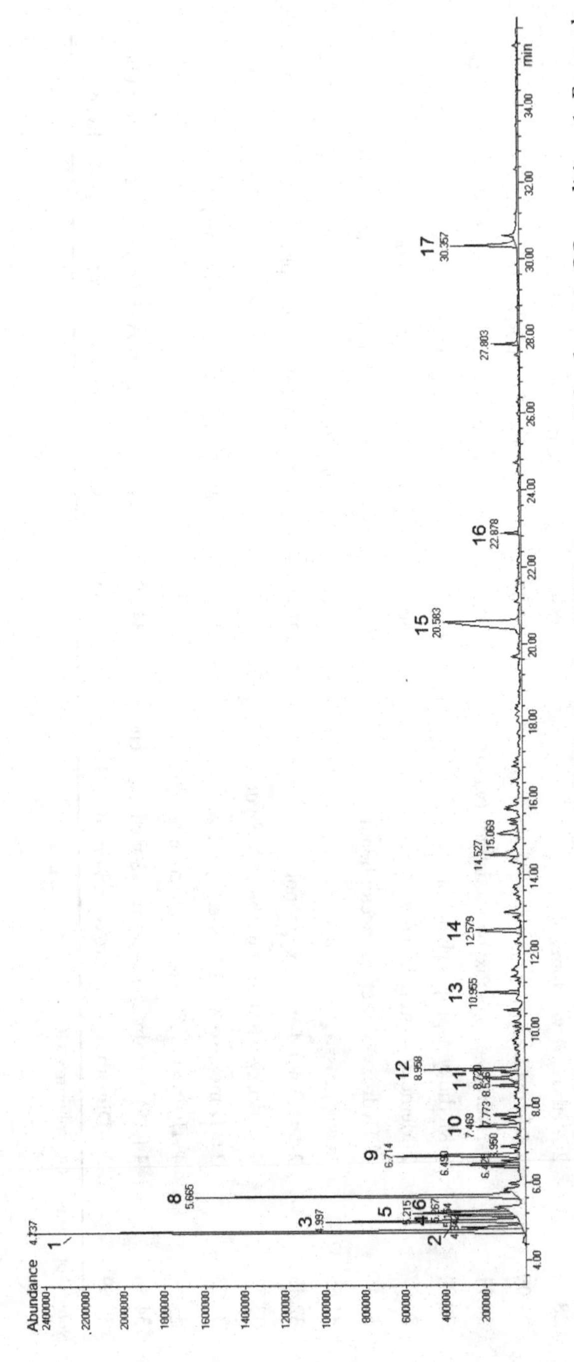

Fig. 5.6. Pyrolysis–GC/MS chromatogram of paper (cellulose fibers) at 550°C. Apparatus 1, GC column 1, GC conditions 1. For peak identification, see Table 5.4.

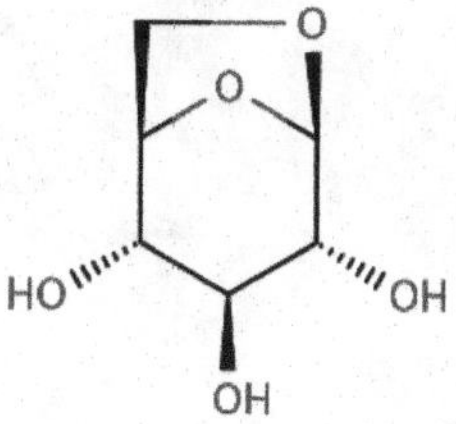

Fig. 5.7. Chemical structure of levoglucosan.

Table 5.4. Pyrolysis products of paper (cellulose fibers) at 550°C.

Peak No.	Retention time t_R (min)	Pyrolysis product
1	4.74	Carbon dioxide
2	4.84	Acetaldehyde
3	5.00	Acetic acid anhydride
4	5.15	Acetic acid
5	5.21	Hydroxyacetaldehyde
6	5.27	2,3-Butanedione
7	5.39	Propylene glycol
8	5.66	1-Hydroxy-2-propanone (Acetal)
9	6.71	1,4-Butanedial (Succinaldehyde)
10	7.47	Furfural
11	8.72	Crotonolactone
12	8.96	2-Hydroxy-2-cyclopenten-1-one
13	10.95	3-Methyl-1,2-cyclopentanedione
14	12.58	Nonanal
15	20.58	Levoglucosan
16	22.88	Palmitic acid
17	30.36	Oleic acid

Notes: Apparatus 1, GC column 1, GC conditions 1.

(acetylated unit). It is made by treating the chitin shells of shrimp and other crustaceans with an alkaline substance, like sodium hydroxide [150]. Chitosan has a number of commercial and possible biomedical uses. It can be used in agriculture as a seed treatment and bio-pesticide, helping

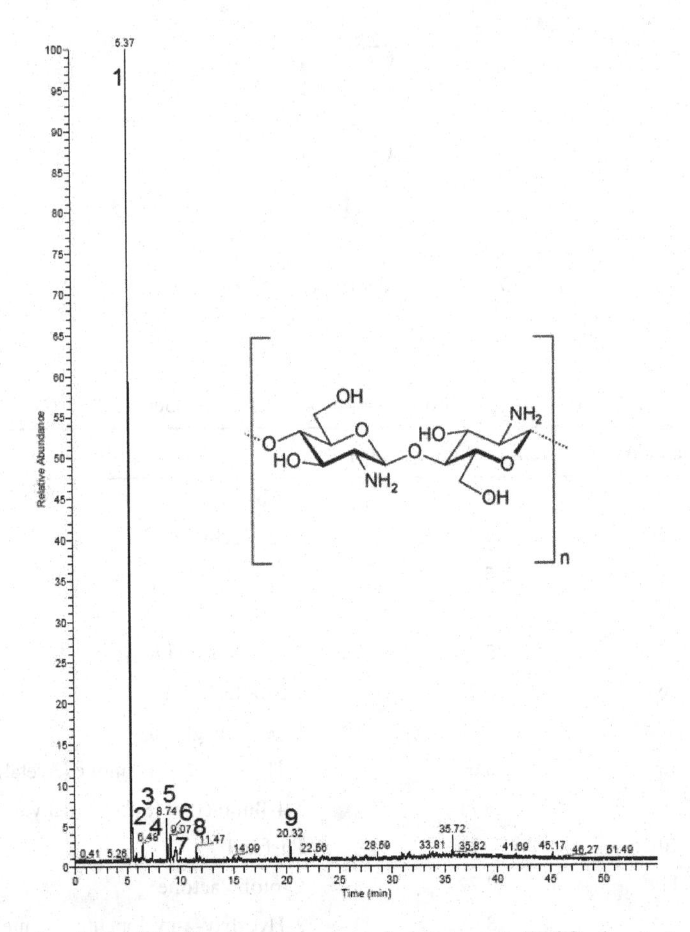

Fig. 5.8. Pyrolysis-GC/MS chromatogram of chitosan at 550°C. Apparatus 2, GC column 3, GC conditions 3. For peak identification, see Table 5.5.

plants to fight off fungal infections. In winemaking, it can be used as a fining agent, also helping to prevent spoilage. In industry, it can be used in a self-healing PU paint coating. In medicine, it may be useful in bandages to reduce bleeding and as an antibacterial agent. It can also be used to help deliver drugs through the skin [150].

Figure 5.8 shows the pyrolysis-GC/MS chromatogram of chitosan at 550°C. The degradation products of chitosan at 550°C are shown in Table 5.5.

Table 5.5. Pyrolysis products of chitosan at 550°C.

Peak No.	Retention time t_R (min)	Pyrolysis product
1	5.37	Carbon dioxide
2	5.54	Acetaldehyde
3	6.48	Acetic formic anhydride
4	7.34	Acetal
5	8.74	Pyrazine
6	9.07	Pyrrole
7	9.59	Acetamide
8	11.47	2-Methylpyrimidine
9	20.32	4-Acetylpyrymidine

Notes: Apparatus 2, GC column 3, GC conditions 3.

5.6. Analysis of Poly(lactic acid)

Polylactide or poly(lactic acid) (PLA) is the front runner in the emerging bioplastics market with the best availability and the most attractive cost structure [151]. The production of the aliphatic polyester from lactic acid, a naturally occurring acid and bulk produced food additive, is relatively straightforward. PLA is a thermoplastic material with rigidity and clarity similar to PS or PET.

PLA can be produced by condensation polymerization directly from its basic building block lactic acid, which is derived by fermentation of sugars from carbohydrate sources, such as corn, sugarcane, or tapioca. Most commercial routes, however, utilize the more efficient conversion of lactide, the cyclic dimer of lactic acid, to PLA via ring-opening polymerization (ROP) catalyzed by a Sn(II)-based catalyst rather than polycondensation. Both polymerization concepts rely on highly concentrated polymer-grade lactic acid of excellent quality [151]. End uses of PLA are in rigid packaging, flexible film packaging, cold drink cups, cutlery, apparel and staple fiber, bottles, injection-molded products, extrusion coating, and so on. PLA is bio-based, resorbable and biodegradable under industrial composting conditions [151].

Figure 5.9 shows the pyrolysis–GC/MS chromatogram of PLA at 550°C and the identification of pyrolysis products of PLA.

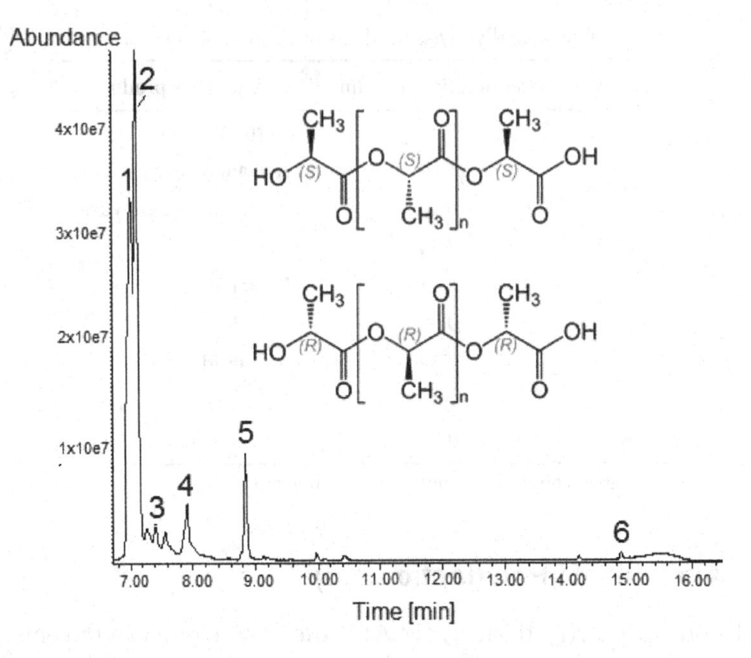

Fig. 5.9. Pyrolysis-GC/MS chromatogram of PLA at 550°C. Apparatus 1, GC column 1, GC conditions 1. Peak identification: (1) t_R = 7.01 min — carbon dioxide, (2) t_R = 7.11 min — acetaldehyde, (3) t_R = 7.41 min — acetic acid, (4) t_R = 7.92 min — acrylic acid, (5) t_R = 8.84 min — toluene (solvent), (6) 14.86 min — lactide (cyclic di-ester of lactic acid).

6

Other Approaches and Developments of the Analytical Pyrolysis

6.1. Reactive Pyrolysis

6.1.1. Thermally assisted hydrolysis and methylation

Modification of the pyrolysis process by introducing chemical reactions has proved to provide additional chemical structure information for various organic materials, which is not readily obtainable by conventional analytical pyrolysis methods. The representative process is pyrolysis in the presence of organic alkali, typically tetramethylammonium hydroxide (TMAH) $[(CH_3)_4NOH]$ or tetramethylammonium acetate (TMAAc) $[(CH_3)_4N(OCOCH_3)]$ [6, 114, 115, 152]. This procedure is also called thermally assisted hydrolysis and methylation (THM), because the samples are hydrolytically decomposed and most of the products are almost simultaneously methylated by 25 wt.% aqueous or methanolic solution of TMAH or 25 wt.% aqueous or methanolic solution of TMAAc. The reactive pyrolysis has been successfully applied to the chemical characterization of synthetic and natural products including resins, lipids, waxes, wood products, soil sediments, and microorganisms [6, 152]. This technique is also very effective for the detailed characterization of the synthetic polymeric materials, especially the condensation polymers such as polyesters (PET, PBT) and PCs.

The mechanism of the hydrolysis/methylation pyrolysis technique was investigated for polyaramids by *in situ* Py–GC [153]. Some parameters that influenced *in situ* methylation by TMAH during pyrolysis were explored. Both pyrolysis temperature and excess TMAH (pH effect) influenced the methylation of carboxyl, aromatic, amino, and hydroxyl functional groups. The solvent for TMAH, methanol, or water significantly affected the methylation for the polyaramids but hardly influenced other model compounds studied. The explanation given assumed a transesterification mechanism rather than hydrolysis/methylation. However, *N*-methylation prior to the decomposition of polyaramids may not be excluded. The reactive pyrolysis technique was also applied to the determination of the sequence distribution of polyacetal copolymers in the presence of cobalt sulfate [154]. Reactive hydrolysis was also practiced to study various kinds of industrially available PCs [6, 152]. Various phenolic compounds were observed in the pyrolysis products. The PC main chain almost quantitatively degraded through reactive pyrolysis at the carbonate linkage to yield the dimethyl derivatives of the constituents, such as bisphenol A, by a hydrolytic pyrolysis reaction in the presence of TMAH. By this method, very accurate and rapid determinations became possible, not only for the chemical composition but also for the level of terminal groups [6, 152].

Compared with the general pyrolysis temperature for the conventional Py–GC or Py–GC/MS (500–700°C), relatively low temperature (300–400°C) is commonly selected for THM–GC or THM–GC/MS to suppress the contribution of random thermal cleavage of the polymer chain at higher temperatures, but high enough for the instantaneous THM reaction. The resulting products are transferred to the separation column and the separated species are detected by a detector. THM with Py–GC/MS (THM Py–GC/MS) results in the methylation of the polar functional groups in the sample (i.e. hydroxyl and carboxylic acid groups). This decreases the polarity of the molecules and increases the chromatographic resolution [155, 156].

THM Py–GC/MS has previously been used for the analysis of a variety of plant samples including whole wood, extracted lignin, waxes, resins, pectins, extractives, cellulose, and other carbohydrates. It has been known for many years that the lignin composition of a sample differs with genus and species. Many authors have found that these differences can be observed using THM Py–GC/MS analysis [155, 156]. During THM Py–GC/MS, the lignin unit linkages are cleaved, and the polar functional groups are methylated, resulting in methoxybenzenes with various side chains. The lignin products can be identified as *p*-hydroxyphenyl, guaiacyl, or syringyl units by observing whether a pyrolysis product is a mono-, di-, or trimethoxyphenyl unit, respectively [155, 156].

Figure 6.1 shows a measuring system for Py–GC in the presence of organic alkali using a microfurnace pyrolyzer along with a typical reaction scheme for an ester and a carbonate linkage [157]. This system is basically the same as that for the ordinary Py–GC. A weighed polymer sample is placed into a sample cup and TMAH reagent is added. The sample cup is then introduced into the pyrolyzer and the polymer sample is instantaneously decomposed in the flow of helium carrier gas. The polymer sample should be grounded into a powder as fine as possible to obtain sufficient reaction efficiency with the organic alkali [157].

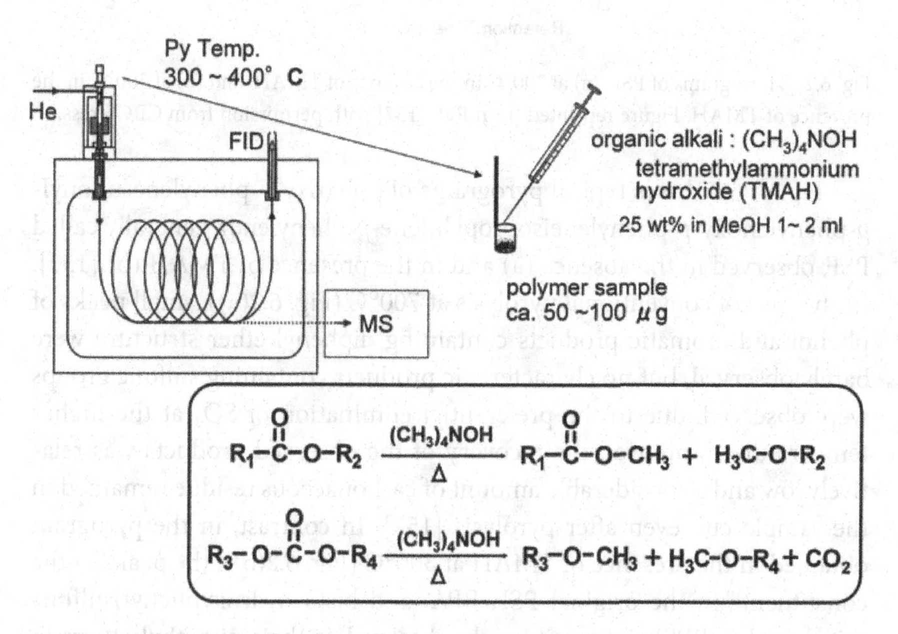

Fig. 6.1. Typical measuring system for Py–GC in the presence of organic alkali. Figure reprinted from Ref. [157] with permission from CRC Press.

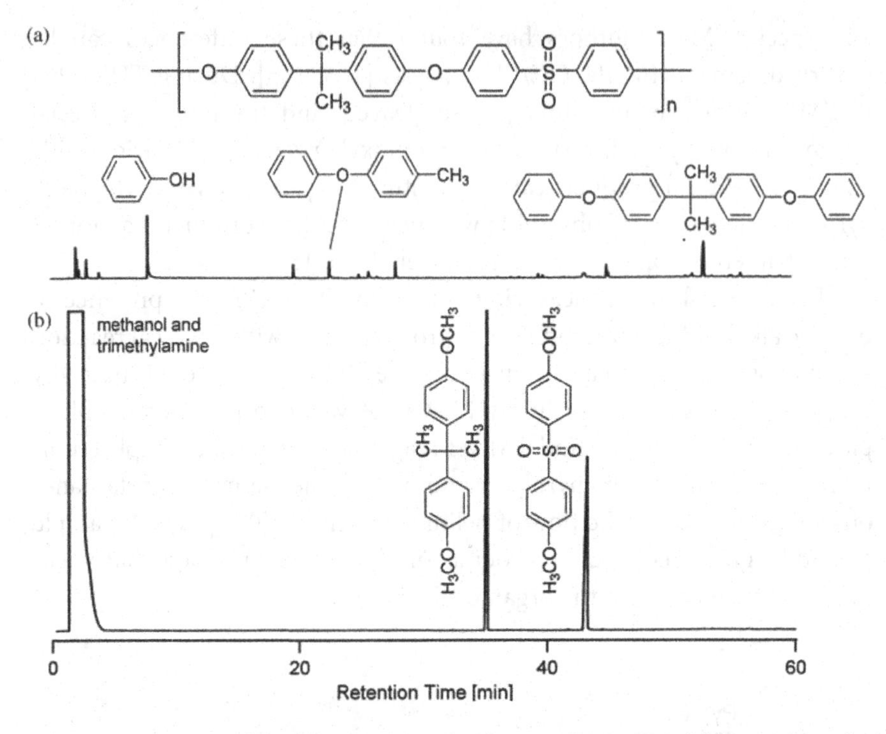

Fig. 6.2. Pyrograms of PSF: (a) at 700°C in the absence of TMAH, and (b) at 300°C in the presence of TMAH. Figure reprinted from Ref. [157] with permission from CRC Press.

Figure 6.2 shows typical pyrograms of poly(oxy-*p*-phenylenesulfonyl-*p*-phenylenoxy-*p*-phenyleneisopropylidene-*p*-phenylene), generally called PSF, observed in the absence (a) and in the presence of TMAH (b) [157]. In the case of conventional pyrolysis at 700°C (Fig. 6.2(a)), small peaks of phenol and aromatic products containing diphenyl ether structure were barely observed, but no characteristic products containing sulfone groups were observed, due to the preferential elimination of SO_2 at the higher temperature. Moreover, the recovery of the observed products was relatively low and a considerable amount of carbonaceous residue remained in the sample cup even after pyrolysis [157]. In contrast, in the pyrogram obtained in the presence of TMAH at 300°C (Fig. 6.2(b)), the peaks of the constituents in the original PSF, BPA, and bis(4-hydroxyphenyl)sulfone (bisphenol S; BPS) were exclusively observed as their dimethyl ethers in almost quantitative recovery without any residues [157].

6.1.2. Silylation

One of the main problems of the pyrolysis technique is the formation of low molecular acidic, alcoholic, and aminic pyrolysis products, which are not really suitable for gas chromatographic analysis, causing a rather low reproducibility of the resulting pyrograms, low sensitivity for specific compounds, and strong memory effects [158]. Moreover, the high fragmentation of natural macromolecules during pyrolysis leads to the formation of many unspecific compounds. To overcome these problems, the sample can be pyrolyzed using a suitable reagent, which transforms polar functionalities into less polar moieties. Thermally assisted reactions using TMAH usually provide good results, because TMAH allows the simultaneous hydrolysis of ester bonds and transmethylation, producing chromatograms whose interpretation is relatively straight forward [158]. Another reactive pyrolysis method is based on *in situ* silylation (derivatization) using hexamethyldisilazane (HMDS) as silylating agent. Silylation is the introduction of a silyl group into a chemical molecule, usually in substitution for active hydrogen in the hydroxyl group (–OH) in alcohols, phenols, carboxylic acids, oximes, sulpho-acids, boric acids, phosphorous acids, in the –NH group in amines, amides, imines, and in the SH group in thiols and thiolcarboxylic acids by a silyl group. Replacement of active hydrogen by a silyl group reduces the polarity of the compound and reduces hydrogen bonding [8]. The silylated derivatives are more volatile and thermally stable. The detection of compounds is enhanced. The trimethylsilyl (TMS) group, $-Si(CH_3)_3$, is the most popular and versatile silyl group for GC and GC/MS analysis. Nonpolar polymethylphenylsiloxane GC stationary phases combine inertness and stability with excellent separating characteristics for these derivatives.

Py–GC/MS with *in situ* derivatization using HMDS as a silylating agent was used for the characterization of the Vietnamese lacquer from *Rhus succedanea* and the Burmese lacquer from *Melanorrhoea usitata* in the work of Colombini's group [158]. Pyrolytic molecular profiles were characterized, MS spectra of pyrolysis products were interpreted, and pyrolytic profiles of silylated alkylcatechols, silylated alkylphenols, aliphatic hydrocarbons and alkylbenzenes, obtained from different lacquers, were compared. The study shows that pyrolysis–GC–MS with *in situ* silylation with HMDS is a suitable analytical approach for the speciation of oriental lacquers.

In another work [159], analyses of samples of acrylic resins used in Fine Arts and Conservation of Cultural Goods have been performed using online trimethylsilylation with HMDS, combined with Py–GC/MS. This method provides suitable signals for compounds corresponding to monomeric, dimer, sesquimer, trimer, and, in some cases tetramer fractions and thus, a large number of compounds were identified by using this proposed technique [159]. On the other hand, the derivatization reagent HMDS has been effective in the derivatization of acrylic and methacrylic groups present in the studied resins, avoiding undesirable broad peak of methacrylic acid (MAA), appearing when direct pyrolysis or online derivatization with TMAH is carried out. The proposed method enables the identification and differentiation of acrylic compounds not only from raw materials but also from historical paintings in which the resin is present at micro or nanoscale and is combined with inorganic materials acting as interferents [159].

Figure 6.3 shows the Py–GC/MS chromatogram obtained from the pyrolysis of MA–EMA (30/70) type copolymer Paraloid B-72 (Rhöm &

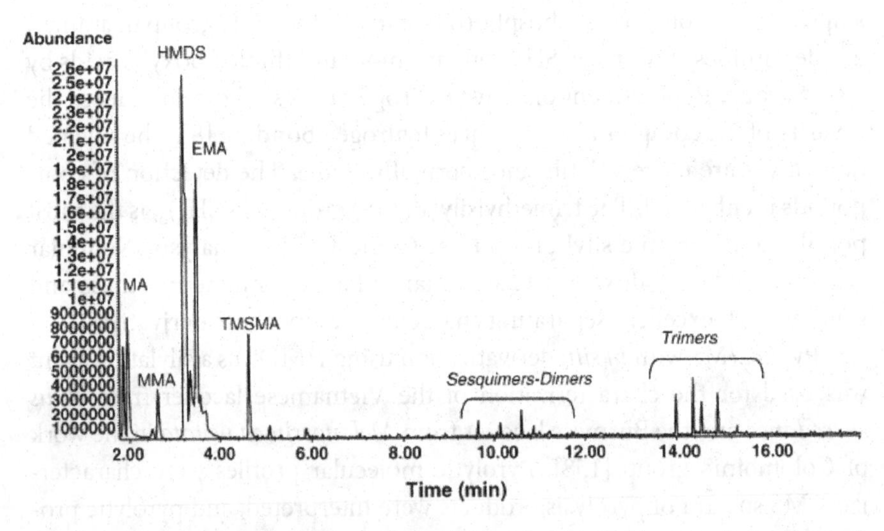

Fig. 6.3. Py–GC/MS chromatogram obtained from the MA–EMA-type polymer Paraloid B72 sample at 700°C by using online trimethylsilylation with HMDS. Peak identification: MA — methyl acrylate, MMA — methyl methacrylate, HMDS — hexamethyldisilazane, EMA — ethyl methacrylate, TMSMA — trimethylsilyl derivative of MAA. Figure reprinted from Ref. [159] with permission from Elsevier.

Haas, Darmstadt, Germany) by using online trimethylsilylation with hexamethylenedisilazane (HMDS). Well-resolved peaks corresponding to MA and EMA were found together with the trimethylsilyl (TMS) derivative of methacrylic acid (TMSMA), whereas acrylic acid peak did not appear. Peak corresponding to MMA also appears in the monomeric region of the pyrogram [159].

In another work [160], Py–GC/MS analyses using thermal-assisted derivatization with the silylating agent 1,1,1,3,3,3-HMDS were performed in art for the elucidation on the composition of original paint materials used by Edvard Munch (1863–1944). The analytical pyrolysis with silylation was used to characterize the gross chemical composition of the paint and to evaluate the presence of macromolecular species like lipid, proteic, or polysaccaridic materials. GC/MS after hydrolysis and derivatization allowed the determination of the fatty acid profile of the paint tubes and to evaluate the molecular changes associated with curing and ageing. Figure 6.4 shows the Py–GC/MS chromatogram of the Winsor & Newton "Mineral Grey" paint sample derivatized with 1,1,1,3,3,3-HMDS.

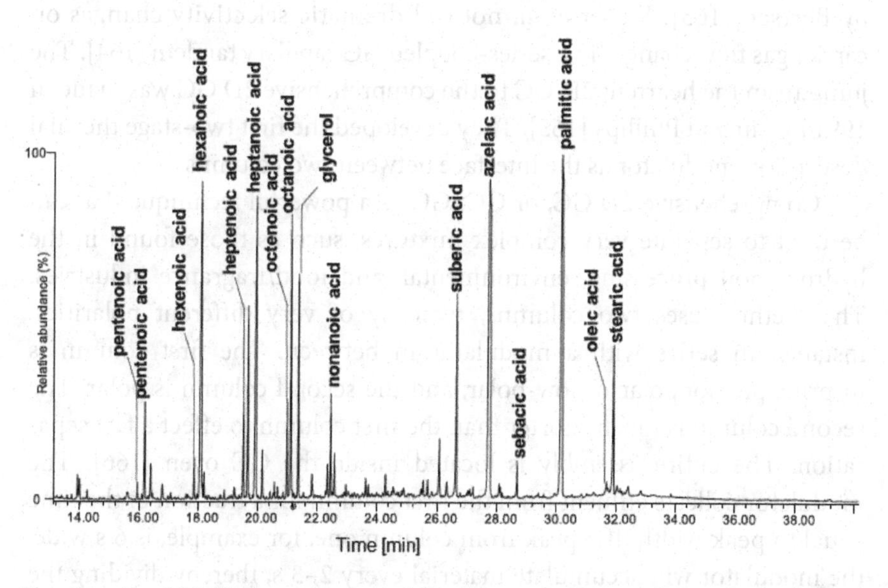

Fig. 6.4. Py–GC/MS chromatogram of the Winsor & Newton "Mineral Grey" paint sample derivatized with 1,1,1,3,3,3-HMDS. Figure reprinted from Ref. [160] with permission from Elsevier.

6.2. Analytical Pyrolysis–Two-Dimensional Gas Chromatography/Mass Spectrometry (Py–GC×GC/MS)

Multidimensional separation techniques involve the sample being dispersed in different time dimensions. When the sample is separated using two dissimilar columns, the total peak capacity will be the product of the peak capacities of the individual columns, representing a huge gain in separation space [161]. The multidimensional GC technique differentiates between the two-dimensional (2D) GC with heartcutting and the comprehensive 2D GC. In the classic heartcut approach, a fraction from the first retention axis is transferred for separation on the second retention axis. By contrast, in the comprehensive approach, after a single introduction to the first column the entire sample is subjected to the two different separations. The earliest work in 2D GC techniques employed valve systems to connect the two columns and for fractionation of the first column eluent. In 1968, Deans introduced a valveless switching system (or stream-switching) using pneumatic pressure balancing [162]. The principle of operation has been reviewed extensively by Bertsch [163]. Kaiser *et al.* noticed dramatic selectivity changes on carrier gas flow changes in a series-coupled GC capillary tandem [164]. The jump from the heartcut 2D GC to the comprehensive 2D GC was made in 1991 by Liu and Phillips [165]. They developed the first two-stage thermal desorption modulator as the interface between two columns.

Comprehensive 2D GC, or GC×GC, is a powerful technique that can be used to separate very complex mixtures, such as those found in the hydrocarbon processing, environmental, and food/fragrance industries. The method uses two columns, typically of very different polarities, installed in series with a modulator in between. The first column is in principle nonpolar or low-polar, and the second column is polar. The second column is much shorter than the first column to effect a fast separation. The entire assembly is located inside the GC oven [166]. The modulator collects effluent from the first column for a fraction of the time equal to peak width. If a peak from column one, for example, is 6 s wide, the modulator will accumulate material every 2–3 s, thereby dividing the peak from the first column into two or three "cuts." The modulator focuses the material collected from each cut into a very narrow band through flow

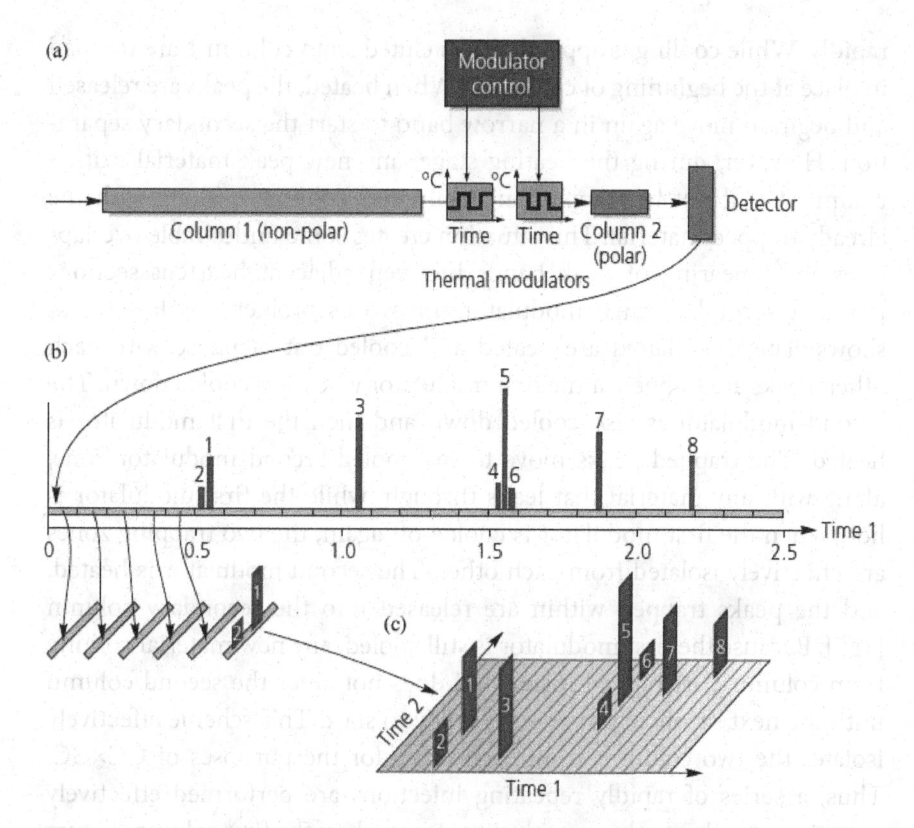

Fig. 6.5. Comprehensive GC×GC system: (a) schematic diagram with dual thermal modulators, (b) representation of a raw GC×GC chromatogram showing positions of individual slices, and (c) representation of a 3D chromatogram reconstructed from the raw chromatogram. Figure and the explanation of the function of comprehensive GC×GC have been taken from Ref. [167].

compression. It introduces the bands sequentially onto the second column, resulting in additional separation for each band injected onto the second column [166].

Figure 6.5(a) illustrates the main principles of comprehensive GC×GC operation with a thermal modulator [167]. The simplest arrangements intermittently blow cooled nitrogen or air over a section of the beginning of column 2. When the cooling flow is turned off, the cooled column section rapidly reheats from circulating hot oven air. Some arrangements positively heat the cooled section to release the trapped peaks more rapidly. While cooling is applied, peaks eluted from column 1 are trapped in place at the beginning of column 2. When heated, the peaks are released and begin to move again in a narrow band to start the secondary separation. However, during the heating stage, any new peak material exiting column 1 will not be trapped and will enter column 2, along with the already trapped material. This situation creates some undesirable overlapping and smearing of peak bands between adjacent heartcut sections [167]. A second thermal modulator solves this problem, as Fig. 6.5(a) shows. The modulators are heated and cooled out of phase with each other. Peaks are trapped at the first modulator when it is cooled down. The second modulator is also cooled down and then the first modulator is heated. The trapped peaks move to the cooled second modulator zone, along with any material that leaks through, while the first modulator is hot. When the first modulator is cooled off again, the two trapping zones are effectively isolated from each other. The second modulator is heated, and the peaks trapped within are released into the secondary column [167]. Because the first modulator is still cooled, any new material coming from column 1 is trapped inside and does not enter the second column until the next secondary analysis is ready to start. This scheme effectively isolates the two columns from each other for the purposes of GC×GC. Thus, a series of rapidly repeating injections are performed effectively onto the secondary column with slices trapped off the first column. Figure 6.5(b) shows a representation of a raw GC×GC chromatogram section that is 2.5 min long. For GC×GC, the chromatogram from the secondary column appears with discrete groups or bands of peaks. Each secondary column run is represented by the 0.1-min-long bands along the baseline. Peaks 1 and 2, which are coeluted from column 1, are separated in the sixth secondary column run. Peaks 4, 5, and 6, which were poorly resolved on column 1, are also separated, but the separation occurs in the 16th secondary column run [167]. The most common data transformation is the construction of a 2D representation in which one axis represents the separation on the first column and the other axis represents the secondary column separation. A contour plot, using elevation lines or color coding, represents the signal intensity [167]. A third dimension can be added optionally by making the *z*-axis proportional to the signal intensity. To generate the reconstructed 2D plot, chromatographers use the timing of the thermal modulators to locate the start of each secondary column run in the raw data stream. The raw data stream is sliced into multiple subchromatograms that represent each of the secondary runs. As Fig. 6.5(c) shows, these slices can be assembled side by side to produce a 2D or 3D map of the overall separation. The map is very useful because it shows relationships between groups of peaks in the primary and the secondary column separations [167].

New visualization and data processing techniques have been developed to display and analyze the 2D retention pattern, and the number of peaks that can be resolved and quantified in the GC×GC chromatogram have been dramatically increased. These advances enable GC×GC to become an excellent technique to analyze complex mixtures, such as natural crude oils and refined petroleum products, or bio-oils. The study of petrochemical composition was one of the earliest applications to benefit from GC×GC, and it remains invaluable today for providing a high degree of component separation and useful information on molecular structure through the well-known "roof-tiling" effect.

New developments in comprehensive 2D gas chromatography (GC×GC) coupled with pyrolysis (Py–GC×GC and Py–GC×GC/MS) offer the prospect of providing more complete and quantitative compositional information of complex organic solids, such as synthetic polymers and copolymers, biopolymers, kerogen, and coals [168].

Figure 6.6(a) shows the conventional Py–GC/MS TIC chromatogram of shale from the Paradox formation at 650°C [168]. The pyrogram illustrates the typical separation and distribution of pyrolyzates from moderately low-sulfur marine shale. This rock contains type-II kerogen that is derived from a mixture of planktonic and microbial organisms with smaller amounts of higher plant matter that was washed into the marine system via rivers [168]. Much of the pyrolysate is poorly resolved and impossible to quantify. Even the major component *n*-alkane/alkene pairs are difficult to measure. Nevertheless, it is very easy to see that the pyrolysis of this source rock yields petroleum with a very different composition than the type-I kerogen deposited in a lake setting. For example, the *n*-alkane/*n*-alkene distribution is shifted toward shorter chain lengths and aromatic hydrocarbons are much more abundant. This difference in product distribution is partially due to the higher thermal maturity of the

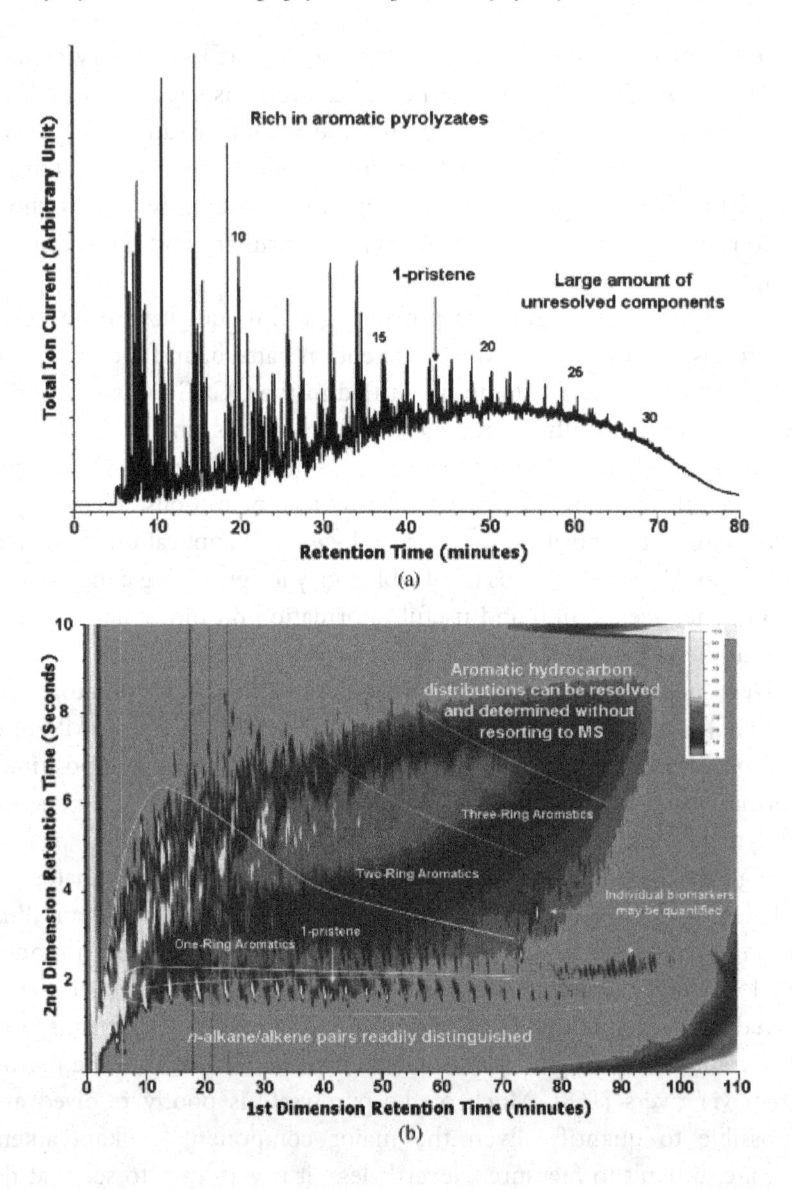

Fig. 6.6. (a) Conventional Py–GC/MS TIC chromatogram at 650°C of marine shale of Paradox formation. The numbers in the figures are numbers of carbons in the molecules. (b) Pyrolysis-comprehensive GC×GC/FID chromatogram at 650°C of marine shale of Paradox formation. Figure reprinted from Ref. [168] with permission from ACS.

Paradox shale but is attributed largely to differences in biotic input. The improved separation using pyrolysis-comprehensive GC×GC-FID of the Paradox formation shale pyrolysate is illustrated in Fig. 6.6(b) [168]. With pyrolysis–GC×GC, the *n*-alkane/alkene pairs are readily determined and easily quantified, as are individual biomarker compounds. Many of the aromatic hydrocarbons with carbon number less than 20 are now individually resolved and can be measured. A substantial portion of the pyrolysate is comprised of aromatic hydrocarbons with carbon number larger than 20 that is still unresolved. Nevertheless, the undifferentiated area can be integrated, providing additional quantitative information [168].

Another application example of the Py–comprehensive GC×GC coupled with TOF-MS for investigation of Murchison meteorite is shown in Fig. 6.7 [169]. As shown in Fig. 6.7, a lot of substances from different classes of organic compounds were well separated and identified by using this technique.

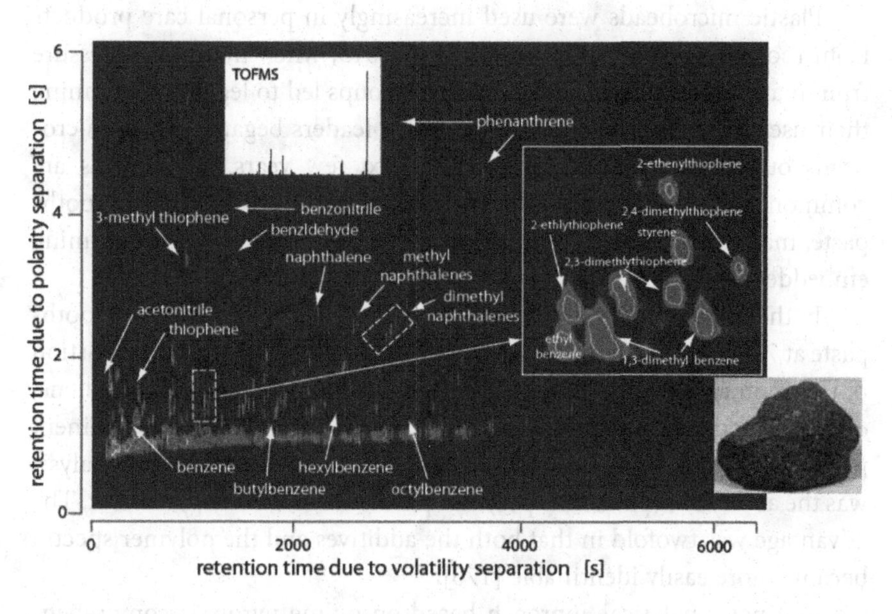

Fig. 6.7. Countour plot from pyrolysis–GC×GC-TOFMS analysis of Murchinson meteorite [169].

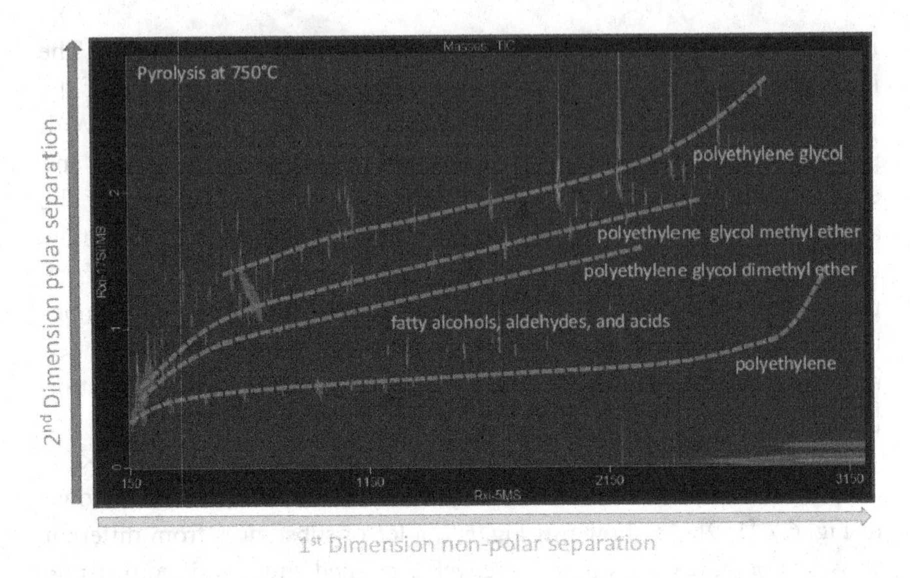

Fig. 6.8. Contour plot from pyrolysis–GC×GC-TOFMS analysis of toothpaste at 750°C [170].

Plastic microbeads were used increasingly in personal care products from toothpaste to facial cleansers until 2013, when mounting pressure from many international environmental groups led to legislation banning their use [170]. As a result, some industry leaders began phasing microbeads out of their products over the next few years. Microbeads are commonly made of PE and other petrochemical-based plastics. In toothpaste, many people have complained about these plastic beads becoming embedded in their gums.

In the application example of the pyrolysis–GC×GC-TOFMS of toothpaste at 750°C, the polymer profile of PE was detected in the lower portion of the contour plot (Fig. 6.8) [170]. PE has a characteristic α,ω-alkadiene, α-alkene, and *n*-alkane profile, with dodecene being the most prominent peak in this instance. The advantage of 2D GC for this polymer analysis was the ability to separate the polymer profile from the sample matrix. The advantage was twofold in that both the additives and the polymer spectra became more easily identifiable [170].

The new analytical approach based on online pyrolysis-comprehensive 2D GC/MS also showed its efficiency in improving the fingerprint of paper samples and revealed its capabilities in distinguishing plant fibers, which currently show similarities in morphological features, and for this reason are indiscernible by microscopic analysis [171]. The investigation was carried out with a multivariate data analysis of the pyrolysis results, using the refined fingerprint issued from the comprehensive 2D GC/MS analyses [171]. This approach confirmed the effectiveness of combining a pyrolysis–GC×GC/MS with multivariate data analysis, namely principal component analysis (PCA) for the differentiation of reference papers, and was extended to the application of museum cases with the characterization of lining papers of unknown origin [171].

6.3 Pyrolysis–Mass Spectrometry (PyMS)

Pyrolysis–mass spectrometry (PyMS) is a sensitive analytical technique in which samples are subjected to rapid thermal degradation (pyrolysis) before the derivative ions are separated in a mass spectrometer. The technique is particularly suitable for the analysis of otherwise nonvolatile compounds in complex samples. The masses of pyrolyzate are expressed as a pyrolysis mass spectrum. This can be interpreted and used to characterize the structures of the parent molecules in the sample, or used in an uninterpreted manner to provide a diagnostic fingerprint that can easily distinguish between related complex samples, such as different microbial strains. Quantitative analysis is possible using sophisticated statistical methods based on neural networks and supervised learning [172].

PyMS is an extremely versatile analytical method, which has been applied in a wide range of areas including organic and inorganic chemistry, synthetic polymer chemistry, natural product chemistry, biochemistry, medicine, pharmacology, microbiology, clinical biology, environmental biology, biotechnology, fuels technology, organic geochemistry, soil science, archeology, forensic science, and food analysis [172]. In geochemistry, PyMS has been used for the analysis of coals, shales, peats, and sediments, but perhaps the most interesting application in this area is in oil exploration, where its ability to detect and characterize volatile hydrocarbons in small pieces of rock, such as drill chips, has greatly enhanced the analysis of cored samples. In archeology, PyMS has been used to determine how ancient artifacts were used. For example, the examination of pottery fragments and other such items has revealed the presence of sealed-in volatile compounds, which allowed the original use of the vessel, from which the fragments came to be determined, e.g. wine jugs, food utensils, and oil containers. In another example, PyMS was used to determine the likely purpose of a 20,000-year-old Neolithic Egyptian grinding stone, which had been buried in the sand since the dawn of history [172]. Many workers have used PyMS to study the structures of polymers, both natural and artificial. Understanding the performance of polymers, in terms of cohesion and substrate adhesion, is of immense commercial significance in the paint and adhesive industries. Similarly, the behavior of polymers under stress and when exposed to external factors such as ultraviolet light has been extensively studied by PyMS, and is useful in the development of novel materials that have desirable properties, e.g. fire-retardant coatings and biodegradable fibers.

At the beginning of the 2000s a Pyrolysis-Evolved Gas Analysis-Mass Spectrometry (Py-EGA-MS) system was developed by Watanabe, Ohtani, Tsuge *et al.* [15–21, 56, 173]. The Py–EGA–MS system using a temperature-programmable furnace-type pyrolyzer (Fig. 6.9), as a heating unit connected to the MS, was successfully applied for the characterization of polymers and copolymers.

In this system, sub-mg of sample is enough to obtain clear EGA thermograms and averaged mass spectra. The specially designed coupling system between the pyrolyzer and MS allows prevent memory effects and condensation of compounds with high boiling points. Py–EGA–MS offers information on thermal properties of samples, similar to the TG and DTG techniques. In Py–EGA–MS, evolved gases formed during the programmed heating of the sample are directly transferred into MS to achieve online monitoring of the components. For the Py–EGA–MS measurements, the separation column in Py–GC/MS is replaced with a deactivated and uncoated stainless steel transfer tube (2 m length, 0.15 mm I.D.), directly connecting a temperature programmable pyrolyzer and an ion source of MS. The resulting thermogram monitored by MS during the programmed heating from 100°C to 700°C at a rate of 20°C/min reflects the evolved gas profile of the sample as a function of temperature. The observed specific pyrograms and/or thermograms and averaged mass spectra provide valuable information regarding the

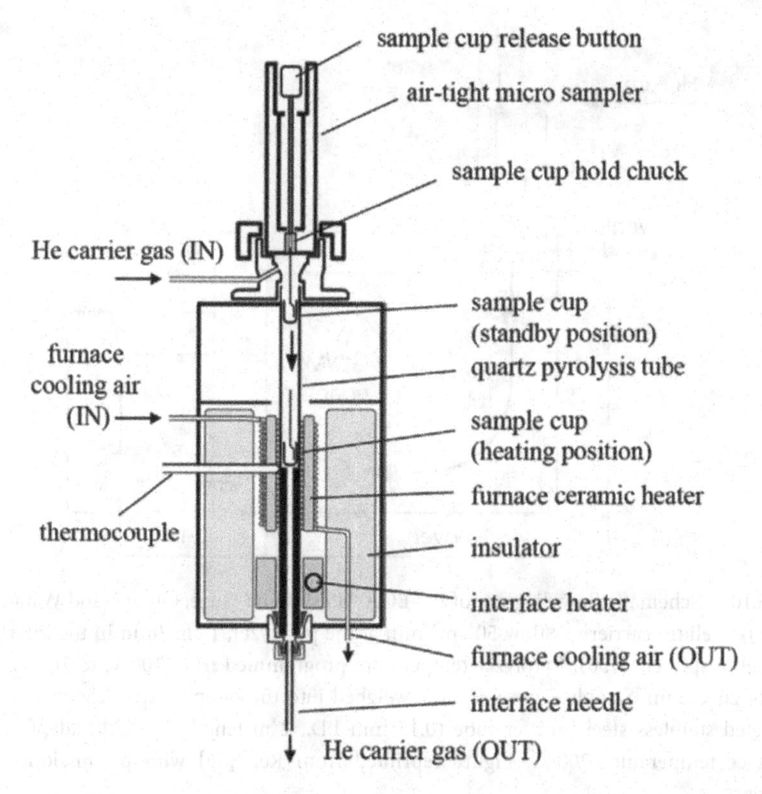

Fig. 6.9. Cross section of temperature programmable furnace-type pyrolyzer. Figure reprinted from Ref. [173] with permission from Elsevier.

composition and/or chemical structures of the original polymer sample, as well as on the degradation mechanisms and related kinetics. Figure 6.10 illustrates the schematic flow diagram of the measuring system of the Py–EGA–MS [56]. In this system, the vertical micro-furnace pyrolyzer mounted on a GC was used under a flow of helium carrier gas. However, any other GC/MS systems equipped with any type of pyrolyzers, such as resistively heated filament devices and inductively heated ones, could provide basically the same tendency data as those in this compilation. Figure 6.11 shows the obtained thermogram data of HDPE by monitoring the TIC of MS, compiled in the upper right corner of the averaged mass spectrum, as a function of the programmed temperature from 100°C to 700°C [56].

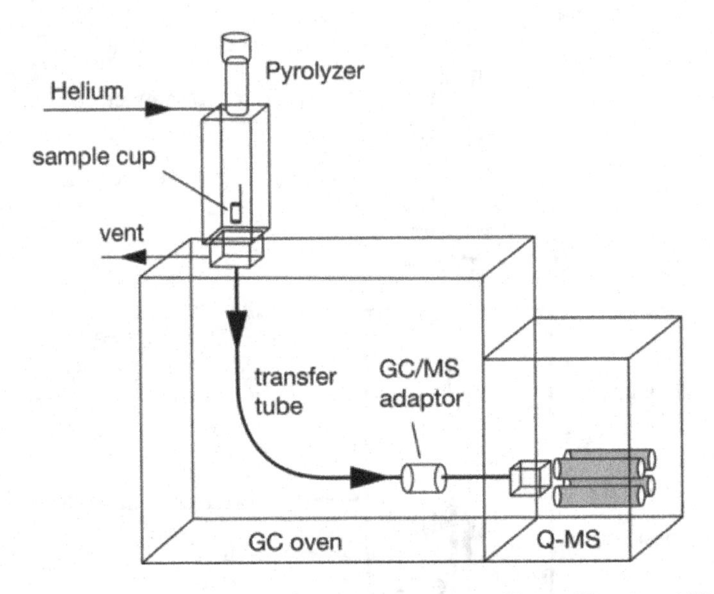

Fig. 6.10. Schematic flow diagram of Py–EGA-MS system (Tsuge, Ohtani, and Watanabe (2011)). Helium carrier gas flow 50 cm^3/min at the pyrolyzer, 1 cm^3/min in the interface through a splitter (1/50). Pyrolyzer temperature programmed from 100°C to 700°C at a rate of 20°C/min. Sample size ca. 0.2 mg weighed into the sample cup. Deactivated and uncoated stainless steel transfer tube (0.15 mm I.D., 2 m length). GC/MS adaptor/MS interface temperature 300°C. Figure reprinted from Ref. [56] with permission from Elsevier.

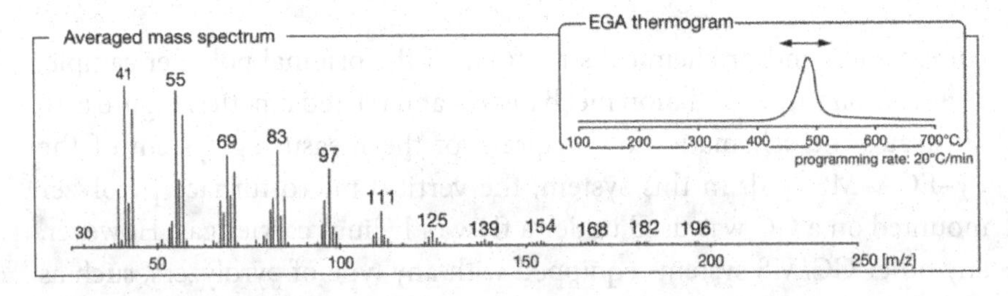

Fig. 6.11. EGA thermogram and averaged mass spectrum of HDPE as a function of the programmed temperature from 100°C to 700°C at a rate of 20°C/min. Sample size ca. 0.2 mg. Figure reprinted from Ref. [56] with permission from Elsevier.

7
Practical Applications of Pyrolysis–GC/MS

The applications of the analytical pyrolysis–gas chromatography/mass spectrometry (Py–GC/MS) range from research and development of new materials, quality control, characterization and competitor product evaluation, medicine, biology and biotechnology, geology, airspace, environmental analysis to forensic purposes or conservation and restoration of cultural heritage. These applications cover analysis and identification of polymers/copolymers and additives in components of automobiles, tires, packaging materials, textile fibres, coatings, adhesives, half-finished products for electronics, paints or varnishes, lacquers, leather, paper or wood products, food, pharmaceuticals, surfactants, and fragrances. In this chapter selected application examples of pyrolysis–GC/MS are described, which originate from the participation of the author (P. K.) in various research projects.

7.1. Application of Py–GC/MS for the Analysis of Microplastics

The global production of plastics in 2014 was already 311 million tons, of which 59 million tons were produced in Europe [96, 174]. At least 270,000 tons of plastic were floating as a giant plastic island on the seas [174]. The waste carpet in the North Pacific (Great Pacific Garbage Patch), which was discovered in 1997, has put together about the size of Germany and France and contains an estimated one million plastic parts per square kilometer [96, 175]. Marine waste can also be found in the Polar region, in deep sea sediments, and on beaches all over the world. During the International Coastal Cleanup in 2013, 648,015 volunteers worked along a stretch of around 13,000 miles of ocean coastline to remove more than 5,500 tons of waste. The top 10 waste items were cigarette butts (913 tons), paper candy wrapping (794 tons), plastic bottles (426 tons), plastic caps and lids (385 tons), plastic soda straws (252 tons), standard plastic shopping bags (200 tons), glass bottles (179 tons), other plastic bags large and small (176 tons), paper bags (167 tons), and beverage cans (154 tons) [96, 176, 177]. Most of the plastic particles are washed into the sea through rivers and eventually sink to the ocean ground. They are carried into the oceans by wind and sewage waters. Larger plastic particles, so-called macroplastics, disintegrate into smaller fragments as a result of UV radiation and wave action in durable microplastics. Plastic particles that are smaller than 5 mm in diameter are called microplasics. Plastics that are manufactured to be of a microscopic size are defined as primary microplastics [96, 178]. These plastics are typically used in facial cleansers and cosmetics or as air-blasting media, while their use in medicine as vectors for drugs is increasingly reported. Under the broader size definitions of a microplastic, virgin plastic production pellets (typically 2–5 mm in diameter) can also be considered as primary microplastics [96, 178].

Microplastics in the sea is a transnational problem that requires an internationally coordinated approach [179]. With the aim of protecting the marine environment, the Marine Strategy Framework Directive (MSFD) was enacted [180]. According to criteria and methodological standards on the good environmental status of marine waters, the "composition of micro-particles (in particular microplastics) has to be characterized in marine litter in the marine and coastal environment" [181, 182]. A critical step following extraction is to identify plastics, especially particles <1 mm in size [183].

Commercial plastics and rubbers always contain low-MW additives. These compounds are essential in polymer/copolymer processing and in the attainment of assuring the end-use properties of a polymer/copolymer. Some of the additives accumulate in the environment and affect our health and the environment. Knowledge of additives is important for evaluating the environmental impact and interaction of polymeric materials to investigate long-term properties and degradation mechanisms [47].

In order to understand the potential impacts of microplastics in the environment, their identification and characterization have been attempted in relation to seawater, sandy and muddy sediments, plankton samples, marine-sewage disposal, sewage effluent, washing-machine effluent, facial cleansers, and vertebrate and invertebrate ingestion [184]. The physico-chemical properties of microplastics, such as size, shape, density, color, and chemical composition, greatly affect their transport in the environment and their bioavailability [184]. The toxicological effects on marine organisms are also influenced by the physico-chemical characteristics of microplastics. Taking this into consideration, the complete characterization of microplastics in terms of abundance, distribution, and chemical composition is of paramount importance in order to characterize their environmental impacts and to enable large-scale spatial and temporal comparisons [184]. The publications of Rocha-Santos and Duarte [184], Hidalgo-Ruz *et al.* [185], and Löder and Gerdts [186] include critical overviews of different analytical approaches used for characterization and quantification of microplastics in the environment. They discussed the most recent studies on their occurrence, fate, and behavior.

7.1.1. Sample processing and characterization of microplastics

Separation of microplastics from samples has been done by direct extraction, density flotation, filtration, or sieving using sieves of several mesh sizes, which allow classification into different size categories [184–186]. After separation of microplastics from water samples, from sediment samples, or from marine biota, the purification of samples is obligatory, especially for instrumental analyses. Biofilms and other organic and inorganic adherents have to be removed from the microplastic particles to avoid artifacts that impede a proper identification [96]. Furthermore, the purification step is necessary to minimize the nonplastic filter residue on filters on which the microplastic fraction <500 μm is concentrated.

The gentle way to clean plastic samples is stirring and rinsing with freshwater. A treatment with 30% hydrogen peroxide of the dried sediment sample, the sample filter or the microplastic particles themselves removes large amounts of natural organic debris [186].

The analysis of microplastics in general can be separated into two major areas: (1) morphological and physical characterization, and (2) chemical characterization and quantification [184]. The characterization of the size distribution of microplastics depends on the sampling and separation methods due to sieve and filter-pore sizes. The size categories can be given using a sieve cascade during the separation procedure or by length measurements assisted by optical microscopy [184]. Scanning electron microscopy (SEM), scanning electron microscopy–energy dispersive X-ray spectroscopy (SEM–EDS), and environmental scanning microscopy–energy dispersive X-ray spectroscopy (ESEM–EDS) were the techniques used for characterizing surface morphology [184]. Quantification of microparticles has been performed to obtain spatial and temporal distribution, to determine rates of accumulation, to investigate the organic pollutants and metals absorbed onto microplastics, and to assess the impacts of microplastics on marine organisms [184].

The identification of chemical composition of microplastics allows for a clear assignment of a sample to a certain polymer/copolymer origin. The common analytical techniques used for the chemical identification of microplastics are the FTIR spectroscopy, the Raman spectroscopy, and the analytical pyrolysis coupled with the GC/MS [184–186]. In these techniques, the identification is based on the comparison of the obtained sample spectrum or pyrogram from the microplastics with the spectra or pyrograms of known polymers/copolymers. Figure 7.1 shows the obtained Py–GC/MS chromatogram of a fisher net microplastics sample pyrolyzed at 700°C. The pyrogram consists of serial triplet-peaks of straight-chain aliphatic C_3–C_{30} hydrocarbons, corresponding to α,ω-alkadienes, α-alkenes, and n-alkanes, respectively, in the order of the increasing $n + 1$ carbon number in the molecule. Such an elution pattern is characteristic for the pyrolysis of PE. Other substances identified in pyrolyzate, like 2-methyl-1-pentene (t_R = 6.17 min) and 2,4-dimethyl-1-heptene (t_R = 8.32 min), indicate the contamination of PE with PP or a blend of both PE and PP.

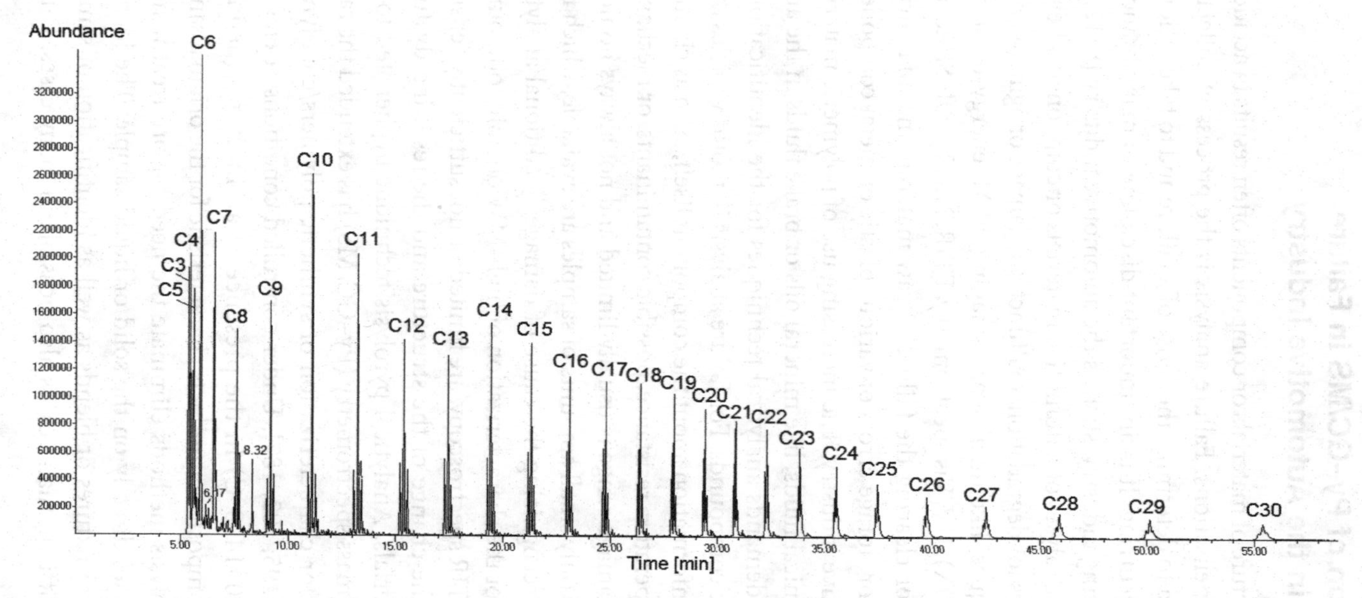

Fig. 7.1. Pyrolysis–GC/MS chromatogram of the investigated fishing net at 700°C. Apparatus 1, GC column 1, GC conditions 2. Peak identification: serial triplet-peaks of straight-chain aliphatic C_3–C_{30} hydrocarbons, corresponding to α,ω-alkadienes, α-alkenes and *n*-alkanes, respectively. Peak at $t_R = 6.17$ min — 2-methyl-1-pentene (propylene dimer), peak at $t_R = 8.32$ min — 2,4-dimethyl-1-heptene (propylene trimer).

7.2. Application of Py–GC/MS in Failure Analysis in the Automotive Industry

Failure of the structure of materials or components often results in accidents and in hefty compensations. Failure analysis is the process of collecting and analyzing data to determine the cause of a failure and to take action to prevent it from recurring. It is an important discipline in many branches of the manufacturing industry, such as the automotive industry [8, 12, 95, 102, 187]. Failure analyses of automotive materials or components help to identify root causes for degradation, malfunction, damage or aging. Various analytical techniques, like microscopy imaging, SEM, energy-dispersive X-ray analysis (EDX), UV/Vis spectrometry, FTIR, NMR, TOF-SIMS, and others are used for clearing the failure of raw material, manufacturing, function, design, or storage errors of various plastic or metal components from the automotive industry. The increasing use of polymeric materials, rubbers, and chemical fluids, like mineral oils or brake fluids in the automotive industry, demands analytical techniques for the identification of high-MW organic compounds. For failure analysis in motor vehicles there is often a lack of information about the component itself, such as chemical composition, temperature resistance, possible contaminants, or mechanical properties. The damage range is usually limited and not always homogeneous. Very often only small amounts of samples are available, which may be important for recognizing the cause of damage. Traditional analytical techniques used for the characterization of high-MW organic compounds, such as TA and FTIR spectroscopy, are limited or not sufficiently sensitive to demonstrate the change of the structure and the resulting dysfunction of used materials. Analytical pyrolysis technique hyphenated to gas chromatography/mass spectrometry (Py–GC/MS) has extended the range of possible tools for characterization of synthetic polymers/copolymers or rubbers [8, 12, 95, 102, 187]. Under controlled conditions at elevated temperature (500–1400°C) in the presence of an inert gas (helium), reproducible decomposition products characteristic for the original sample are formed. Pyrolysis methods eliminate the need for pre-treatment by performing analyses directly on the solid or liquid sample. The identification of complex mixtures or blends as well as identification of samples with so-called "difficult matrices" is also possible in many cases. Due to these small sample amounts, the investigation of heterogeneous polymers with a coarse phase or a gradient composition structure (phase separation, poor mixing, etc.) is sophisticated and may lead to great variations in the measuring results. In this case a multiple determination of different positions of the investigated part is essential to achieve a significant image of its composition [102].

7.2.1. Identification of fouling on the wall of a hydraulic cylinder from the automotive industry

Figure 7.2 shows the fouling on the wall of a hydraulic cylinder from the automotive industry, which failed by a leak between upper and lower pressure chambers. The question was, from which components (seals, pressure-, or test media) of the piston was the resulting fouling. The sampling was

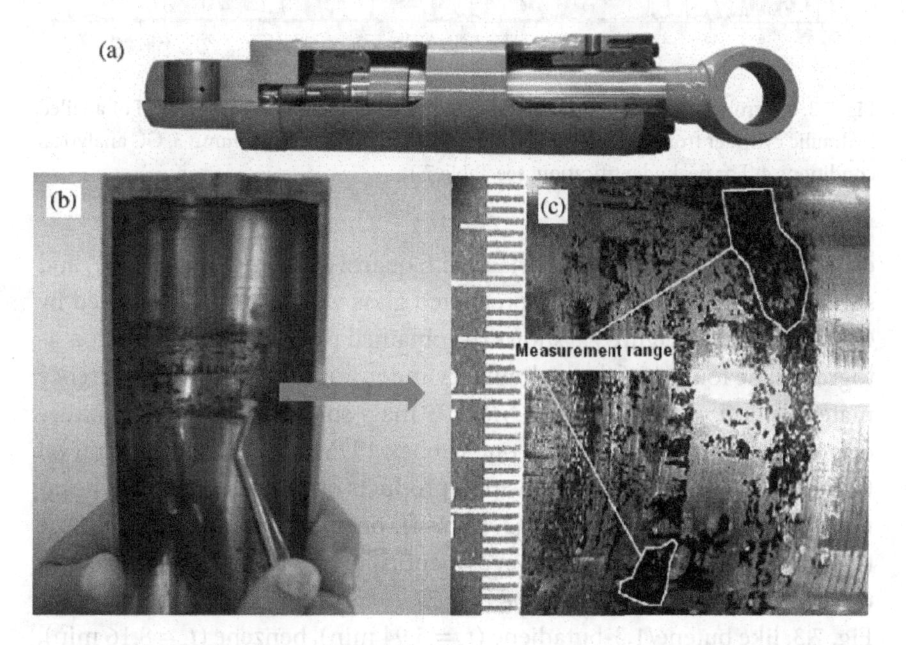

Fig. 7.2. Fouling on the wall of a failed hydraulic cylinder: (a) general view of a hydraulic cylinder from the automotive industry, (b) sampling of fouling by rubbing the affected surface with the quartz glass wool, (c) zoom of the measurement range. Figure reprinted from Ref. [95] with permission from Elsevier.

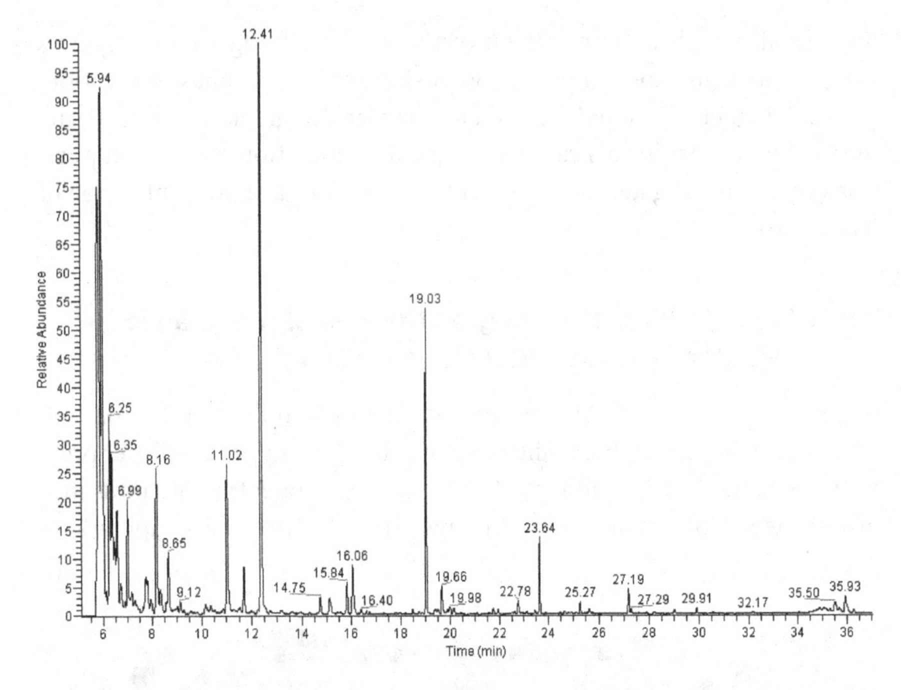

Fig. 7.3. Pyrolysis–GC/MS chromatogram at 700°C of the fouling on the wall of a failed hydraulic cylinder from the automotive industry. Apparatus 2, GC column 3, GC analytical conditions 4. For peak identification, see Table 7.1.

done by rubbing the affected surface with quartz glass wool (Fig. 7.2(b)) and subsequent pyrolysis of the impregnated glass wool at 700°C, followed by GC/MS analysis. Figure 7.3 shows the obtained pyrolysis–GC/MS chromatogram. The fouling of the hydraulic cylinder was identified with reference materials and with help of the *NIST 05* mass spectra library as a mixture of poly(dimethylsiloxane) (silicone rubber, PDMS), SBR, and mineral oil (motor oil). The main decomposition products of silicone rubber are hexamethylocyclotrisiloxane (t_R = 12.41 min), octamethylocyclotetrasiloxane (t_R = 19.03 min), decamethylocyclopentasiloxane (t_R = 23.64 min), and dodecamethylocyclohexasiloxane (t_R = 27.19 min) (Fig. 7.3). Other peaks in Fig. 7.3, like butene/1,3-butadiene (t_R = 5.94 min), benzene (t_R = 8.16 min), toluene (t_R = 11.02 min), and styrene (t_R = 16.06 min), are typical pyrolysis products of SBR. Originally suspected hydrogenated nitrile-butadiene rubber (HNBR), from which the sealing rings were made, was excluded.

Table 7.1. Pyrolysis products at 700°C of the fouling on the wall of a failed hydraulic cylinder from the automotive industry.

Retention time t_R (min)	Pyrolysis product	Pyrolyzed material
5.94	Butene/1,3-butadiene	SBR
6.25	1-Pentene	Mineral oil
8.16	Benzene	SBR
8.65	1-Heptene	Mineral oil
11.02	Toluene	SBR
11.72	1-Octene	Mineral oil
12.41	Hexamethylocyclotrisiloxane	PDMS
15.84	1-Nonene	Mineral oil
16.06	Styrene	SBR
19.03	Octamethylocyclotetrasiloxane	PDMS
19.66	1-Decene	Mineral oil
22.78	1-Undecene	Mineral oil
23.64	Decamethylocyclopentasiloxane	PDMS
27.19	Dodecamethylocyclohexasiloxane	PDMS
34–36.5	Heavy *n*-alkanes	Mineral oil

Notes: Apparatus 2, GC column 3, GC analytical conditions 4.

In the pyrolyzate of the fouling there were not identified pyrolysis products of HNBR, such as methacrylonitrile, aniline, benzonitrile, or tolunitrile. The presence of *n*-alkenes and heavy *n*-alkanes in pyrolyzate is characteristic for the pyrolysis of mineral oil (motor oil) (Table 7.1).

The results obtained show that a lot of information about dysfunction of automobile parts can be obtained from the fouling material on the surface of the failed parts. In such cases the sampling can be made by rubbing the affected surface with the quartz glass wool, followed by the pyrolysis–GC/MS of the enriched wool.

7.2.2. Identification of fouling on the internal surface of a bearing race from a car

In the following case, the composition of fouling on the internal surface of a bearing race from a car was to identify (Fig. 7.4). The sampling was done by

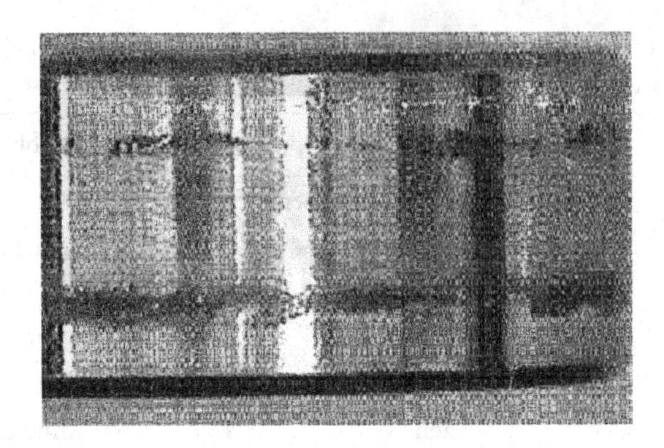

Fig. 7.4. Fouling on the internal surface of a bearing race from a car.

rubbing the affected surface with quartz glass wool and subsequent pyrolysis of the impregnated glass wool at 700°C, followed by GC/MS analysis.

Figure 7.5 shows the obtained pyrolysis–GC/MS chromatogram at 700°C. The fouling of the bearing race was identified with reference materials and with the help of the *NIST 05* mass spectra library as a mixture of mineral oil, SBR, polyamide 6.6 (PA 6.6, Nylon), and epoxy resin. The pyrolysis products of the investigated fouling as well as their origin are summarized in Table 7.2. The presence of *n*-alkenes C_3H_6 (propene) up to $C_{20}H_{40}$ (1-eicosene) in pyrolysate is characteristic for the pyrolysis of mineral oil (motor oil). The peaks of butene/1,3-butadiene, benzene, toluene, ethylbenzene, *o*-xylene, styrene, and α-methylstyrene indicate the presence of SBR. The peak of cyclopentanone (retention time t_R = 10.46 min) is characteristic for the pyrolysis of PA 6.6 (Nylon), while the peaks of phenol and *p*-isopropenylphenol could be generated from the epoxy resin (Fig. 7.5 and Table 7.2). The identified materials could then be assigned to the corresponding car components. The obtained analytical results could then be used for troubleshooting with the customer.

7.2.3. Identification of membranes from hydraulic cylinders from the automotive industry

In the next case it was to clarify, why two rubber membranes from hydraulic cylinders from the automotive industry show poor properties and are

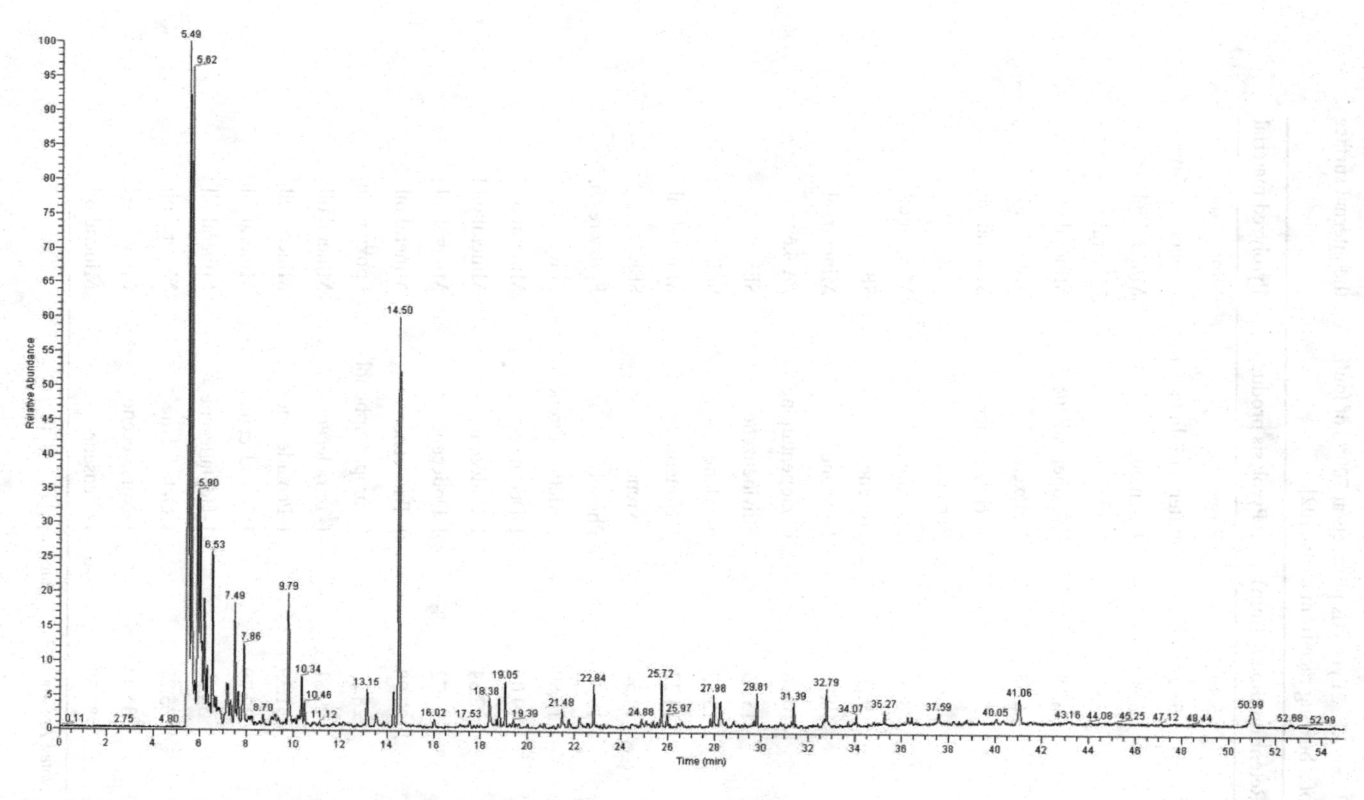

Fig. 7.5. Pyrolysis–GC/MS chromatogram at 700°C of fouling on the internal surface of a bearing race from a car [102]. Apparatus 2, GC column 3, GC analytical conditions 3.

Table 7.2. Pyrolysis products at 700°C of fouling on the internal surface of a bearing race from a car [102].

Retention time t_R (min)	Pyrolysis product	Pyrolyzed material
5.49	Propylene	Mineral oil
5.62	Butene/1,3-butadiene	Mineral oil, SBR
5.90	1-Pentene	Mineral oil
5.99	Pentadiene	Mineral oil
6.18	Cyclopentadiene	Mineral oil
6.53	1-Hexene	Mineral oil
7.14	Cyclohexadiene	Mineral oil
7.49	Benzene	SBR
8.70	1-Heptene	Mineral oil
9.79	Toluene	SBR
10.34	1-Octene	Mineral oil
10.46	Cyclopentanone	PA 6.6
13.15	Ethylbenzene	SBR
13.52	*o*-Xylene	SBR
14.27	1-Nonene	Mineral oil
14.50	Styrene	SBR
18.38	Phenol	Epoxy resin
18.80	α-Methylstyrene	SBR
19.05	1-Decene	Mineral oil
22.84	1-Undecene	Mineral oil
25.72	1-Dodecene	Mineral oil
27.98	1-Tridecene	Mineral oil
28.25	Isopropenylphenol	Epoxy resin
29.81	1-Tetradecene	Mineral oil
31.39	1-Pentadecene	Mineral oil
32.79	1-Hexadecene	Mineral oil
34.07	1-Heptadecene	Mineral oil
35.27	1-Octadecene	Mineral oil
36.44	1-Nonadecene	Mineral oil
37.59	1-Eicosene	Mineral oil

Notes: Apparatus 2, GC column 3, GC analytical conditions 3.

Fig. 7.6. Investigated rubber membrane from the hydraulic cylinder from the automotive industry.

cracked. We have analyzed the well- and the poor-functioning membranes. Figure 7.6 shows the photo of a membrane from the hydraulic cylinder from the automotive industry.

The membranes were identified as poly(acrylonitrile-*co*-1,3-butadiene) (nitrile rubber, NBR) based on the decomposition products like 1,3-butadiene, acrylonitrile, methacrylonitrile, benzene, toluene, styrene, benzonitrile, and tolunitrile (Fig. 7.7 and Table 7.3). The main feature of pyrolysis of nitrile rubber is the formation of the monomers 1,3-butadiene (peak 1, t_R = 7.16 min) and acrylonitrile (peak 2, t_R = 7.39 min). The presence of benzonitrile (peak 7, t_R = 12.26 min) in pyrograms is also characteristic for the pyrolysis of nitrile rubber (see Section 4.6.5). Other substances that appear in Fig. 7.7(c) (peaks 11–13) have been identified as thermal degradation products of the nitrile rubber additives. The identified benzothiazole (peak 11, t_R = 16.94 min) has been formed by the thermal degradation of 2-mercaptobenzothiazole (MBT). 2-Mercaptobenzothiazole is used as accelerator for the vulcanization of rubber and as an antioxidant. *N*-Phenyl-1,4-benzenediamine (peak 13, t_R = 28.97 min) was generated from *N*-(1-methylethyl)-*N*-phenyl-1,4-benzenediamine during the pyrolysis of the

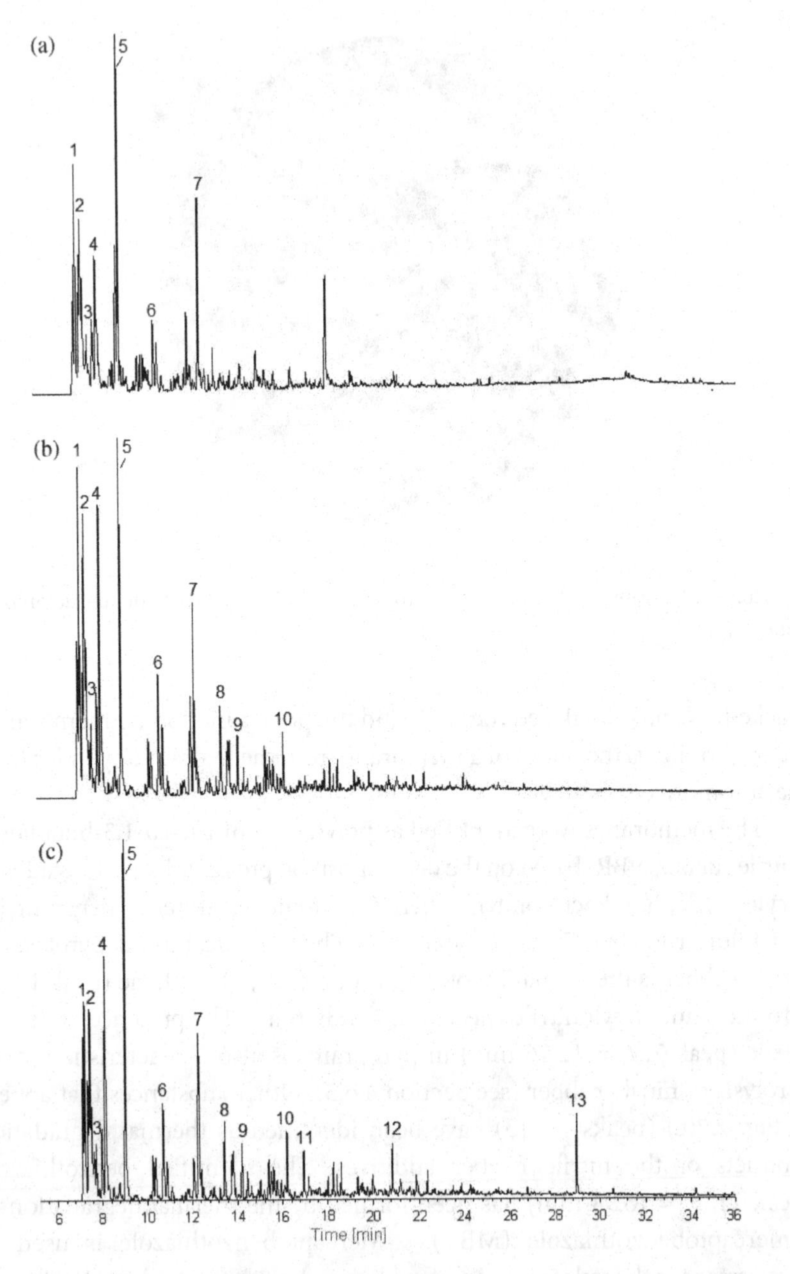

Fig. 7.7. Pyrolysis–GC/MS chromatograms at 700°C of the membranes from hydraulic cylinders from the automotive industry: (a, b) poor-functioning membranes, (c) well-functioning membrane. Apparatus 1, GC column 1, GC conditions 1.

Table 7.3. Pyrolysis products at 700°C of the membranes from the hydraulic cylinders from the automotive industry.

Peak No.	Retention time t_R (min)	Pyrolysis product
1	7.16	1,3-Butadiene
2	7.39	Acrylonitrile
3	7.66	Methacrylonitrile
4	8.07	Benzene
5	8.99	Toluene
6	10.66	Styrene
7	12.26	Benzonitrile
8	13.43	Indene
9	14.17	Tolunitrile
10	16.19	Naphthalene
11	16.94	Benzothiazole
12	20.84	2,4-Dimethylquinoline
13	28.97	*N*-Phenyl-1,4-benzenediamine

Notes: Apparatus 1, GC column 1, GC conditions 1.

NBR sample. This substance is a very effective antioxidant and antiozonant, which provides medium- to long-term protection for all synthetic and natural rubbers. The additives were not identified in the two membranes in Figs. 7.7(a) and 7.7(b) with poor properties.

7.3. Applications of Py–GC/MS to Solve Technological Problems in the Industry

7.3.1. Determination of the crosslinking degree of SDVB copolymers

The chemical composition of a copolymer material has a substantial impact on its properties and performance. Defining a composition profile is therefore an essential aspect of copolymer product development. Since copolymer composition often varies considerably with MW, a detailed analysis is required in copolymer R&D [188].

Traditional polymer characterization techniques such as SEC, coupled with differential refractive index detection (SEC-dRI) or NMR, are either insufficient to perform a comprehensive analysis, or to involve significant labor, complexity, and/or cost. A common characterization technique for copolymer analysis is proton nuclear magnetic resonance (^{1}H NMR), which excels at determining chemical identity. Standard NMR is only able to provide measurements of the sample-average composition. ^{1}H NMR can be used with preparative fractionation to obtain a coarse size/composition distribution, but this may be impractical to perform regularly [188].

This example demonstrates the application of the pyrolysis–GC/MS method for the quantitative determination of crosslinking degree of SDVB copolymers. In order to determine the degree of crosslinking of the copolymers, the samples were pyrolyzed at 700°C and analyzed by GC/MS (Fig. 7.8, Table 7.4). The peak areas in pyrograms obtained were then used for calculating the peak area ratio divinylbenzene/styrene. The value of the crosslinking degree of the copolymer was calculated based on the obtained values of the peak area ratio divinylbenzene/styrene.

The pyrogram in Fig. 7.8 shows an example in which the peak area ratio divinylbenzene/styrene was equal to 0.0151. By the result, the value of the crosslinking of the SDVB sample was determined as 1.5%.

7.3.2. Identification of two industrial plastic batches

In this application example the analysis of two plastic batches from the industry is presented. Both of the plastic batches were pyrolyzed at 700°C, analyzed by GC/MS, and identified by using of the *NIST 05* mass spectra library.

The pyrogram of the first batch is presented in Fig. 7.9(a). The sample was identified as a blend of PE and EBA copolymer. In the pyrogram of the sample (Fig. 7.9(a)) a serial of triplets corresponding to C_3–C_{30} α,ω-alkadienes, α-alkenes, and n-alkanes, respectively, in the order of increasing $n + 1$ carbon number in the molecule, characteristic for the pyrolysis of PE were identified. 1-Butanol detected at $t_R = 6.51$ min indicates the presence of EBA. The small signals of benzene ($t_R = 6.64$ min), toluene ($t_R = 7.63$ min), and styrene ($t_R = 9.53$ min) (Fig. 7.9(a)) indicate the presence of additives.

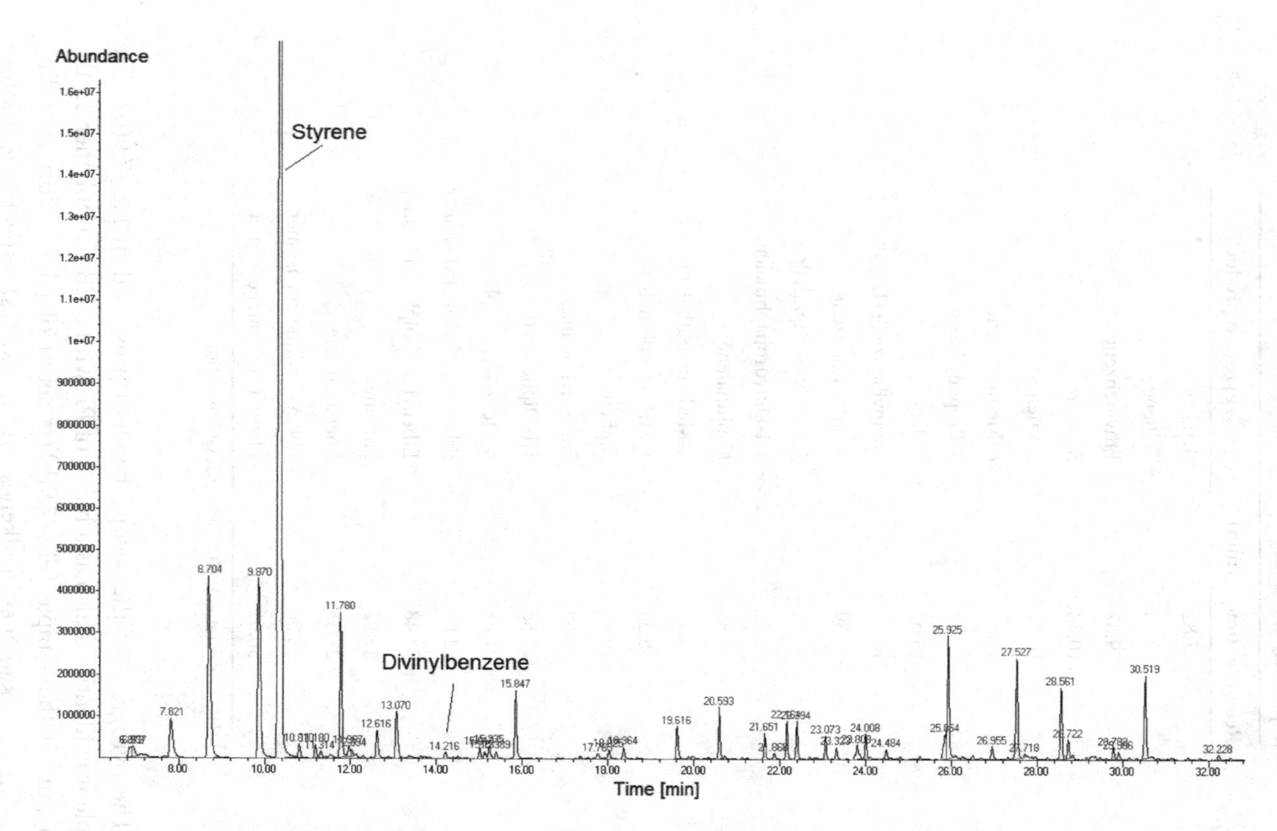

Fig. 7.8. Pyrolysis–GC/MS chromatogram of the investigated SDVB copolymer at 700°C. Apparatus 1, GC column 2, GC conditions 1.

Table 7.4. Pyrolysis products of the investigated SDVB copolymer at 700°C.

Retention time t_R (min)	Pyrolysis product
7.82	Benzene
8.70	Toluene
9.87	Ethylbenzene
10.38	Styrene
10.81	Cumene
11.78	α-Methylstyrene
12.62	2-Propenylbenzene
13.07	Indene
14.22	Divinylbenzene (DVB)
14.39	1-Butenylbenzene
15.01	1,2-Dihydronaphthalene
15.13	1,4-Dihydronaphthalene
15.85	Naphthalene
18.02	2-Methylnaphthalene
18.36	1-Methylnaphthalene
19.62	Biphenyl
20.59	Diphenylmethane
21.65	1,1-Diphenylethane
21.86	4-Methyl-1,1′-biphenyl
24.00	Ethenyl-1,1′-biphenyl isomer
24.48	Ethenyl-1,1′-biphenyl isomer
25.92	Stilbene
27.53	Phenanthrene
28.56	Phenylnaphthalene isomer
30.52	Phenylnaphthalene isomer

Notes: Apparatus 1, GC column 2, GC conditions 1.

The pyrogram of the second batch is presented in Fig. 7.9(b). The sample was identified also as a blend of PE with a EBA copolymer. In the pyrogram of the sample (Fig. 7.9(b)) a serial of triplets corresponding to C_3–C_{30} α,ω-alkadienes, α-alkenes, and *n*-alkanes, respectively, in order

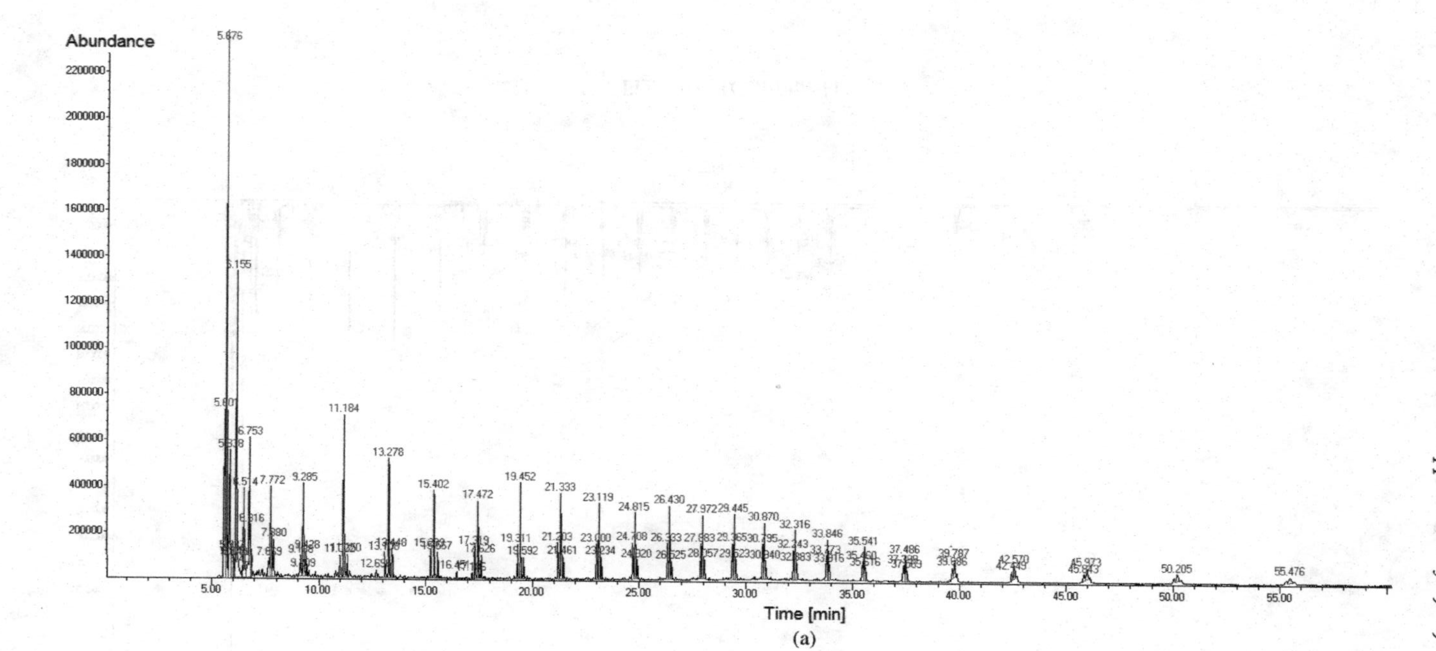

Fig. 7.9. Pyrolysis–GC/MS chromatograms of two industrial plastic batches at 700°C. Apparatus 1, GC column 1, GC conditions 2. For peak identification, see text.

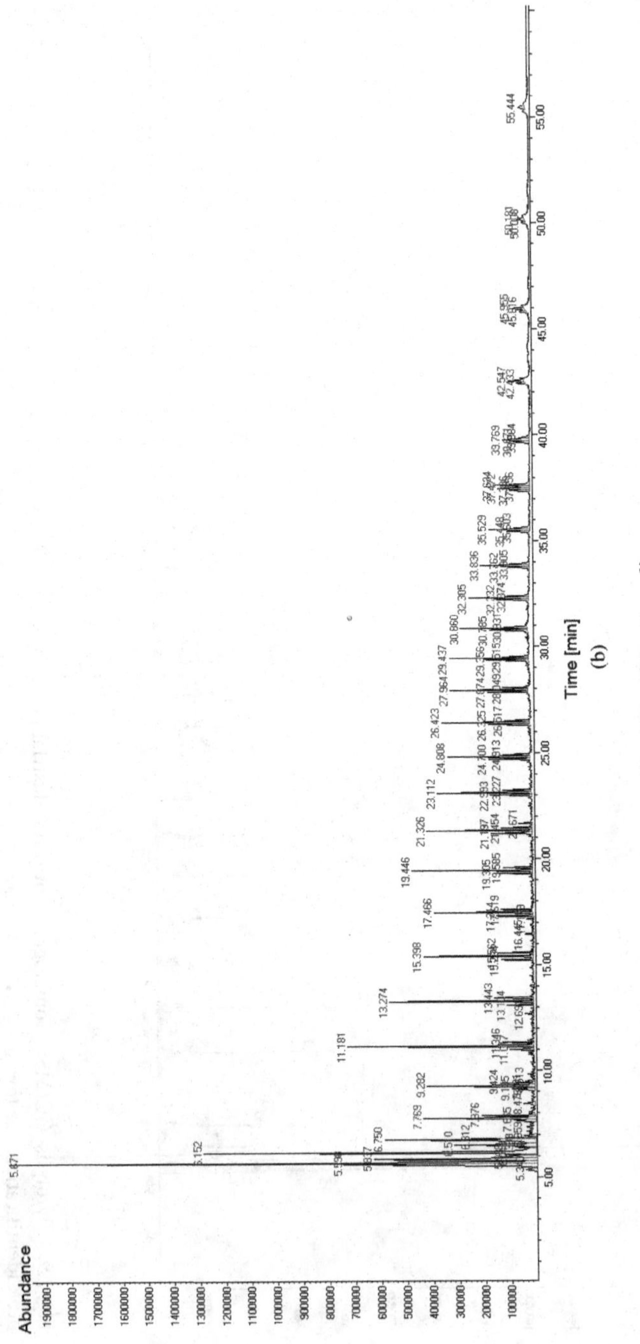

Fig. 7.9. *(Continued)*

of increasing *n*+1 carbon number in the molecule, characteristic for the pyrolysis of PE were identified. 1-Butanol (t_R = 6.51 min) indicates the presence of EBA. Additionally, PP was determined in the second batch. 2,4-Dimethyl-1-heptene (propylene trimer, t_R = 8.47 min) is characteristic for the pyrolysis of PP. PP was not identified in the firs batch (Fig. 7.9(a)). The small signals of benzene (t_R = 6.64 min), toluene (t_R = 7.63 min), and styrene (t_R = 9.53 min) (Fig. 7.9(b)) indicate the presence of additives.

7.3.3. Investigation of an inhomogeneous industrial rubber component

Synthetic elastomers have replaced NR (polyisoprene) to an astonishing degree and account for more than 70% of the rubber used today. In the United States alone, 5 million tons of synthetic rubbers are produced annually. The principal synthetic rubber elastomer is a copolymer of butadiene and styrene. The latex form of rubber and synthetic elastomers has applications in carpet and gloves. Coagulated latex is used for the production of tires and mechanical goods [189]. Much of the polybutadiene rubber produced is blended with polyisoprene to give it improved resilience and lower rolling resistance. More than half of all usage is in tires. Other applications are footwear, wire and cable insulation, and conveyor belts.

Figure 7.10 shows the pyrolysis–GC/MS chromatograms of two places of an inhomogeneous industrial rubber component at 700°C. The identified pyrolysis products, by using the *NIST 05* mass spectra library, are summarized in Table 7.5. As can be seen from Fig. 7.10 and from Table 7.5, the samples examined differ from each other. The degradation products of the sample presented in. Fig 7.10(a) and in Table 7.5 allow the identification of the rubber as pure polyisoprene (IR) without additives. The second rubber sample was identified as a blend of polybutadiene (butadiene rubber, BR) and polyisoprene with additives, like *N*-isopropyl-*N*′-phenyl-*p*-phenylenediamine (Antioxidant 4010 NA) and *N*,*N*′-bis(1,4-dimethylpentyl)-*p*-phenylenediamine (Antioxidant 4030) (Fig. 7.10(b) and Table 7.5). *N*-Isopropyl-*N*′-phenyl-*p*-phenylenediamine is a widely used antioxidant with high efficiency and many purposes, especially for NR, many kinds of synthetic rubber products and their latexes. *N*,*N*′-Bis(1,4-dimethylpentyl)-*p*-phenylenediamine (Antioxidant 4030) is used as an antiozonant in processing of diene rubbers.

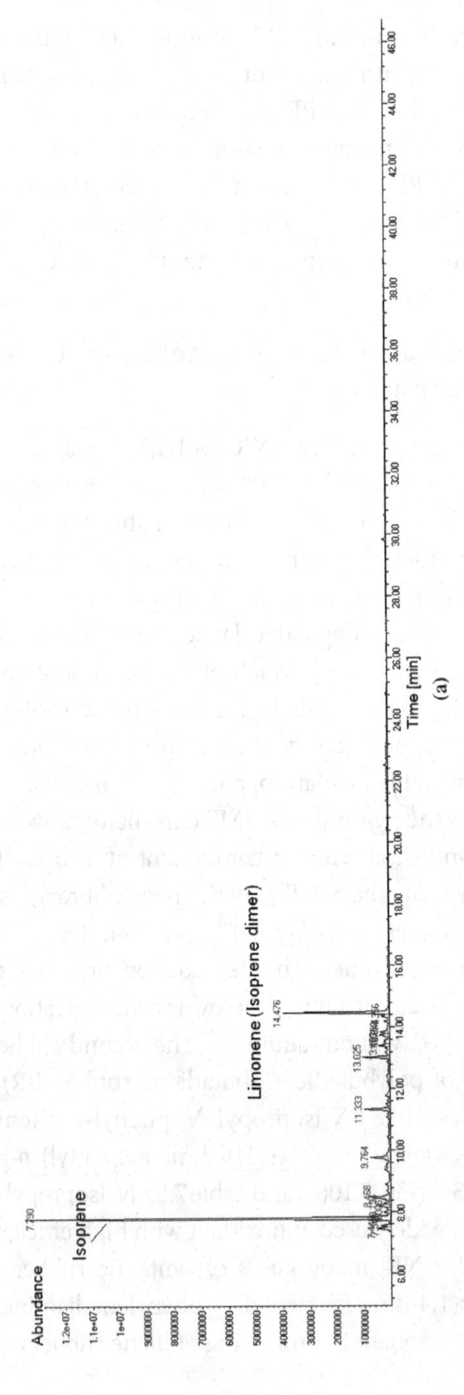

Fig. 7.10. Pyrolysis–GC/MS chromatograms of two places (a and b) of an inhomogeneous industrial rubber component at 700°C. Apparatus 1, GC column 1 UI, GC conditions 2.

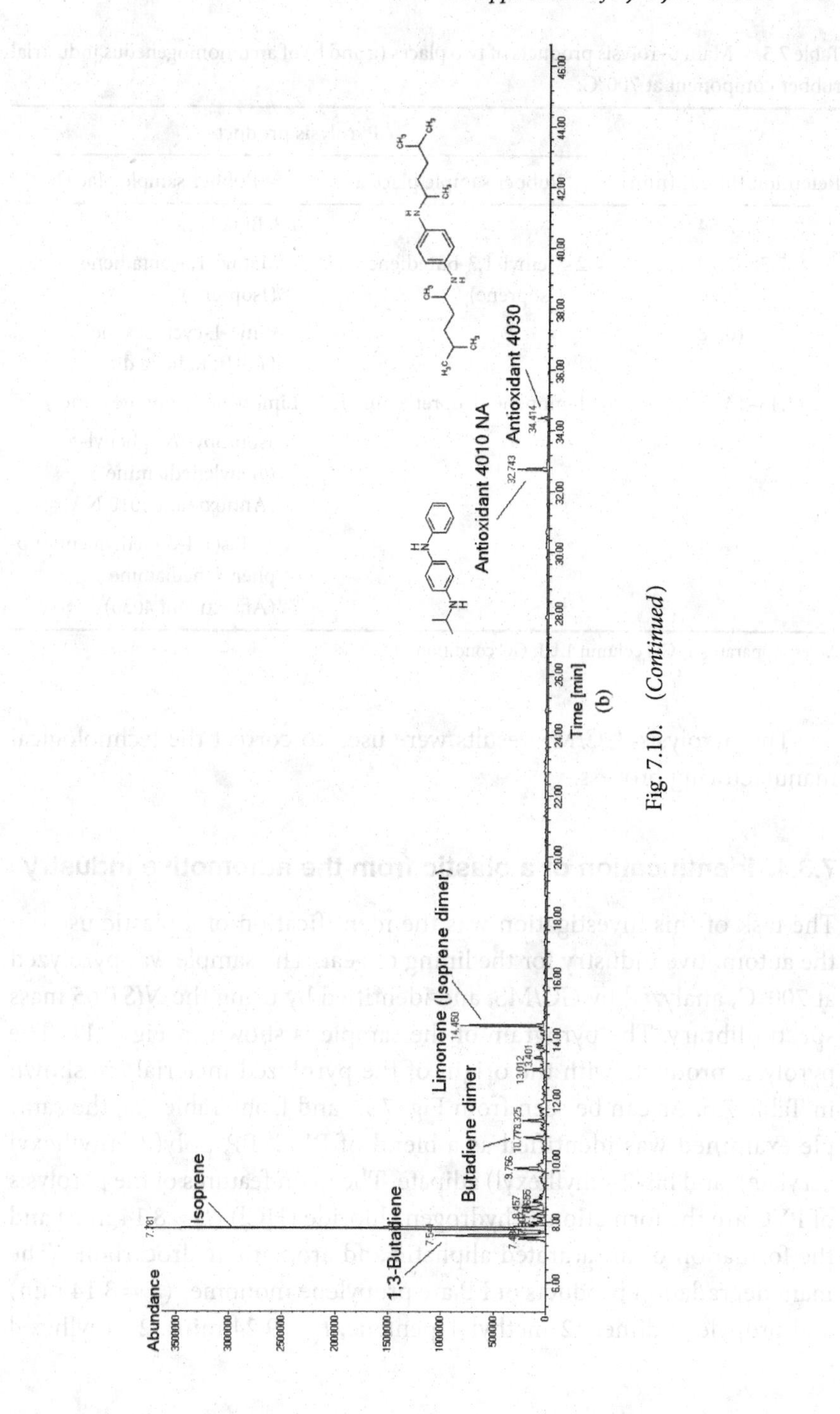

Fig. 7.10. *(Continued)*

Table 7.5. Main pyrolysis products of two places (a and b) of an inhomogeneous industrial rubber component at 700°C.

	Pyrolysis product	
Retention time t_R (min)	Rubber sample place a	Rubber sample place b
7.54		1,3-Butadiene
7.78–7.79	2-Methyl-1,3-butadiene (Isoprene)	2-Methyl-1,3-butadiene (Isoprene)
10.78		4-Vinyl-1-cyclohexene (1,3-Butadiene dimer)
14.45–14.48	Limonene (Isoprene dimer)	Limonene (Isoprene dimer)
32.74		*N*-Isopropyl-*N*′-phenyl-*p*-phenylenediamine (Antioxidant 4010 NA)
34.41		*N*,*N*′-Bis(1,4-dimethylpentyl)-*p*-phenylenediamine (Antioxidant 4030)

Notes: Apparatus 1, GC column 1 UI, GC conditions 2.

The pyrolysis–GC/MS results were used to correct the technological manufacturing process.

7.3.4. Identification of a plastic from the automotive industry

The task of this investigation was the identification of a plastic used in the automotive industry for the lining of gear. The sample was pyrolyzed at 700°C, analyzed by GC/MS, and identified by using the *NIST 05* mass spectra library. The pyrogram of the sample is shown in Fig. 7.11. The pyrolysis products with the origin of the pyrolyzed material are shown in Table 7.6. As can be seen from Fig. 7.11 and from Table 7.6, the sample examined was identified as a blend of PVC, PP, poly(2-ethylhexyl acrylate), and bis(2-ethylhexyl) adipate. The main features of the pyrolysis of PVC are the formation of hydrogen chloride (HCl) ($t_R = 8.14$ min) and the formation of unsaturated aliphatic and aromatic hydrocarbons. The main degradation products of PP are propylene monomer ($t_R = 8.14$ min) and propylene dimer (2-methyl-1-pentene, $t_R = 9.24$ min). 2-Ethylhexyl

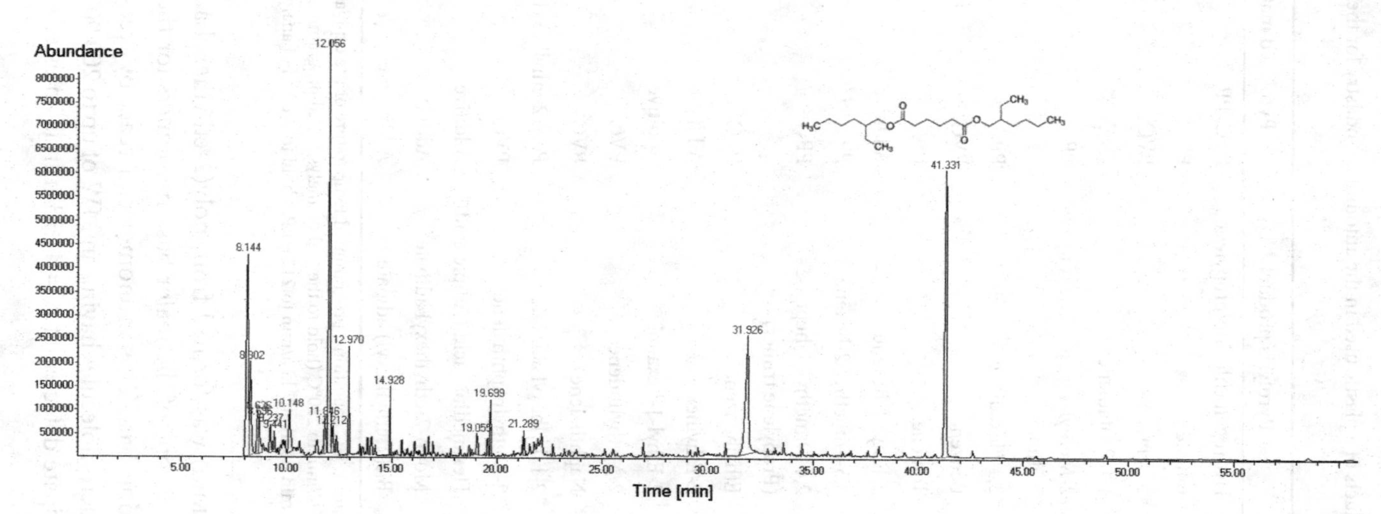

Fig. 7.11. Pyrolysis–GC/MS chromatogram of a plastic used in the automotive industry for the lining of gear at 700°C. Apparatus 1, GC column 1, GC conditions: programmed temperature of the capillary column from 60°C (1-min hold) at 7°C/min to 280°C (hold to the end of analysis) and programmed pressure of helium from 122.2 kPa (1-min hold) at 7 kPa/min to 212.9 kPa (hold to the end of analysis).

Table 7.6. Pyrolysis products of a plastic used in the automotive industry for the lining of gear at 700°C.

Retention time t_R (min)	Pyrolysis product	Pyrolyzed material
8.14	Hydrogen chloride/propylene	PVC, PP
8.30	Butene	PVC
8.64	2-Pentene	PVC
8.70	1,4-Pentadiene	PVC
9.24	2-Methyl-1-pentene (Propylene dimer)	PP
9.44	2,4-Hexadiene	PP
10.15	Benzene	PVC
11.85	Toluene	PVC
12.06	2-Ethyl-1-hexene	Additive
12.40	3,5-Dimethyl-2-hexene	PP
12.97	2,4-Dimethyl-1-heptene (Propylene trimer)	PP
13.51	Ethylbenzene	PVC
14.93	2-Ethylhexanal	Additive
16.09	2-Ethyl-1-hexanol	Additive
18.69	2-Methylindene	PVC
19.56	Naphthalene	PVC
19.70	2-Ethylhexyl acrylate	Poly(2-ethylhexyl acrylate)
21.77	2-Methylnaphthalene	PVC
21.90	Hexanedioic acid (Adipic acid)	Additive
31.93	Mono-2-ethylhexyl adipate	Additive
41.33	Bis(2-ethylhexyl) adipate	Additive

Notes: Apparatus 1, GC column 1, GC conditions: programmed temperature of the capillary column from 60°C (1-min hold) at 7°C/min to 280°C (hold to the end of analysis) and programmed pressure of helium from 122.2 kPa (1-min hold) at 7 kPa/min to 212.9 kPa (hold to the end of analysis).

acrylate (t_R = 19.70 min) was formed from poly(2-ethylhexyl acrylate). 2-Ethylhexyl acrylate is one of the major base monomers for the preparation of acrylate adhesives. This monomer can react by free-radical polymerization to macromolecules having an MW of up to 200,000 g/mol. Bis(2-ethylhexyl) adipate detected at t_R = 41.33 min is an ester of adipic acid and 2-ethyl-1-hexanol. It is a plasticizer and finds application in plastic technology, in order to impart flexibility to rigid polymers. It is an indirect food additive formed due to the contact of adhesives with polymers. Other substances like 2-ethyl-1-hexene, 2-ethylhexanal, 2-ethyl-1-hexanol, hexanedioic acid (adipic acid), and mono-2-ethylhexyl adipate detected in pyrolyzate of sample have arisen during the pyrolysis of the additive bis(2-ethylhexyl) adipate.

8
Conclusions

Analytical Py–GC/MS proved to be a valuable hyphenated technique for the analysis and identification of synthetic organic polymeric materials, their additives and for the analysis of biopolymers. This technique allows the direct analysis of very small sample amounts (5–200 μg) without the need of time-consuming sample preparation or after a simple sample derivatization. Py–GC/MS can be applied to research and development of new materials, quality control, characterization and competitor product evaluation, medicine, biology and biotechnology, geology, airspace, and environmental analysis to forensic purposes or conservation and restoration of cultural heritage. These applications cover analysis and identification of polymers/copolymers and additives in components of automobiles, tires, packaging materials, textile fibres, coatings, adhesives, half-finished products for electronics, paints, or varnishes, lacquers, leather, paper, or wood products, food, pharmaceuticals, surfactants, and fragrances. Pyrolysis–GC/MS allows the confirmation of the source of a failed product, the identification of contaminants causing failure, the competitive analysis, as well as overcoming a problem in product development or quality control. The obtained analytical results are then used for troubleshooting and remedial action of the technological process. This technique could also be used to study the chemical composition of microplastics in the environmental analytics.

References

[1] P. Kusch, G. Knupp, A. Morrisson, Analysis of synthetic polymers and copolymers by pyrolysis–gas chromatography/mass spectrometry, in *Horizons in Polymer Research*, R. K. Bregg (Ed.), Nova Science Publishers Inc., New York, USA, 2005, pp. 141–191.

[2] M. A. Kruge, Analytical pyrolysis principles and applications to environmental science, in *Environmental Application of Instrumental Chemical Analysis*, M. M. Barbooti (Ed.), Apple Academic Press Inc., Waretown, NJ, USA, 2015, pp. 533–567.

[3] P. C. Uden, Nomenclature and terminology for analytical pyrolysis, IUPAC Recommendations 1993, *Pure Appl. Chem.*, 65(11), 1993, 2405–2409.

[4] B. Bruce Sitholé, Pyrolysis in the pulp and paper industry, in *Encyclopedia of Analytical Chemistry*, R. A. Meyers (Ed.), John Wiley & Sons Ltd., Chichester, UK, 2000, pp. 8443–8481.

[5] R. E. Majors, *Sample Preparation Fundamentals for Chromatography*, Agilent Technologies Inc., Wilmington, DE, USA, 2013.

[6] S. Tsuge, H. Ohtani, Pyrolysis-gas chromatography, in *Practical Gas Chromatography*, K. Dettmer-Wilde, W. Engewald (Eds.), Springer-Verlag, Berlin, Heidelberg, Germany, 2014, pp. 829–847.

[7] S. Tsuge, Analytical pyrolysis — past, present and future, *J. Anal. Appl. Pyrol.*, 32, 1995, 1–6.

[8] P. Kusch, D. Schroeder-Obst, V. Obst, G. Knupp, W. Fink, J. Steinhaus, Application of pyrolysis–gas chromatography/mass spectrometry (Py-GC/MS) and scanning electron microscopy (SEM) in failure analysis for the identification of organic compounds in chemical, rubber and automotive industry,

in *Handbook of Materials Failure Analysis with Case Studies from the Chemicals, Concrete and Power Industries*, A. S. Hamdy Makhlouf, M. Aliofkhazraei (Eds.), Elsevier, Amsterdam, The Netherlands, 2016, pp. 441–469.

[9] D. O. Hummel, Mass spectrometry, in *Atlas of Plastics Additives, Analysis by Spectrometric Methods*, D. O. Hummel (Ed.), Springer, Berlin, Heilderberg, Germany, 2002, pp. 73–109.

[10] D. O. Hummel, F. Scholl, Py-GC gekoppelt mit Massenspektrometrie, in *Atlas der Polymer- und Kunststoffanalyse*, D. O. Hummel, F. Scholl (Eds.), Hanser Verlag, München, Germany, 1988.

[11] M. L. Hallensleben, H. Wurm, Polymeranalytik, *Nachr. Chem. Techn. Lab.*, 37(5), 1989, M1–M52.

[12] P. Kusch, Pyrolysis–gas chromatography/mass spectrometry of polymeric materials, in *Advanced Gas Chromatography — Progress in Agricultural, Biomedical and Industrial Applications*, M. A. Mohd (Ed.), InTech, Rijeka, Croatia, 2012, pp. 343–362.

[13] T. Wampler (Ed.), *Applied Pyrolysis Handbook*, 2nd ed., CRC Press, Boca Raton, USA, 2007.

[14] http://www.sge.com/products/gc-accessories/pyrojector-ii accessed Jan 3, 2017.

[15] www.frontier-lab.com accessed Jan 3, 2017.

[16] S. Tsuge, H. Ohtani, C. Watanabe, Y. Kawahara, Application of a multifunctional pyrolyzer for evolved gas analysis and pyrolysis–GC of various synthetic and natural materials, *Am. Lab.*, 35(1), 2003, 32–35.

[17] S. Tsuge, H. Ohtani, C. Watanabe, Y. Kawahara, Application of a multifunctional pyrolyzer for evolved gas analysis and pyrolysis–GC of various synthetic and natural materials; part 2, *Am. Lab.*, 35(3), 2003, 48–52.

[18] S. Tsuge, H. Ohtani, C. Watanabe, Application of a multifunctional pyrolyzer for evolved gas analysis and pyrolysis–GC of various synthetic and natural materials; part 3, *Am. Lab.*, 35(12), 2003, 16–18.

[19] S. Tsuge, H. Ohtani, C. Watanabe, Application of a multifunctional pyrolyzer for evolved gas analysis and pyrolysis–GC of various synthetic and natural materials; part 4, *Am. Lab.*, 36(2), 2004, 22–26.

[20] C. Watanabe, A. Hosaka, Y. Kawahara, P. Tobias, H. Ohtani, S. Tsuge, GC–MS analysis of heart-cut fractions during evolved gas analysis of polymeric materials, *LCGC North Am.*, 20(4), 2002, 374–378.

[21] http://www.frontier-lab.com/catalog/en/AS-1020E_E.pdf accessed Jan 3, 2017.

[22] http://www.gerstel.com/pdf/s00135-035-02-Pyrolyzer_en.pdf accessed Jan 3, 2017.

[23] W. Simon, H. Giacobbo, Thermische Fragmentierung und Strukturbestimmung organischer Verbindungen, *Chemie-Ingenieur-Technik*, 37, 1965, 709–714.

[24] H.-R. Schulten, W. G. Fischer, H. J. Wallstab, New automatic sampler for Curie-point pyrolysis in combination with gas chromatography, *J. High Resolut. Chromatogr. Chromatogr. Commun.*, 10(8), 1987, 467–469.

[25] P. Kusch, Application of the Curie-point pyrolysis-high resolution gas chromatography for analysis of synthetic polymers and copolymers, *Chem. Anal. (Warsaw)*, 41, 1996, 241–252.

[26] W. G. Fischer, P. Kusch, Curie-Punkt Pyrolyse und Gaschromatographie. Neue Anwendungsmöglichkeiten für chemische Prüfverfahren (in German), *Materialprüfung*, 33, 1991, 193–195.

[27] W. G. Fischer, P. Kusch, Automatic sampler for Curie-point pyrolysis-gas chromatography with on-column introduction of pyrolysates, *J. Chromatogr.*, 518, 1990, 9–19.

[28] http://www.gsg-analytical.com/english/pyrolyse.htm accessed Jan 5, 2014.

[29] http://www.jai.co.jp/english/products/py/pg387.html accessed Jan 5, 2017.

[30] S. C. Moldoveanu, *Analytical Pyrolysis of Synthetic Organic Polymers*, Elsevier, Amsterdam, The Netherlands, 2005.

[31] http://www.cdsanalytical.com/instruments/pyrolysis/pyroprobe-5000.html accessed Jan 12, 2017.

[32] I. Tydén-Ericsson, A new pyrolyzer with improved control of pyrolysis conditions, *Chromatographia*, 6(8), 1973, 353–358.

[33] E. M. Andersson, I. Ericsson, Determination of the temperature — time profile of the sample in pyrolysis–gas chromatography, *J. Anal. Appl. Pyrol.*, 1(1), 1979, 27–38.

[34] I. Ericsson, Sequential pyrolysis–gas chromatographic study of the decomposition kinetics of *cis*-1,4-polybutadiene, *J. Chromatogr. Sci.*, 16(8), 1978, 340–344.

[35] I. Ericsson, Influence of pyrolysis parameters on results in pyrolysis-gas chromatography, *J. Anal. Appl. Pyrol.*, 8, 1985, 73–86.

[36] http://www.pyrolab.se accessed Jan 11, 2017.

[37] N. K. Meruva, L. A. Metz, S. R. Goode, S. L. Morgan, UV laser pyrolysis fast gas chromatography/time-of-flight mass spectrometry for rapid characterization of synthetic polymers: Instrument development, *J. Anal. Appl. Pyrol.*, 71(1), 2004, 313–325.

[38] L. A. Metz, N. K. Meruva, S. L. Morgan, S. R. Goode, UV laser pyrolysis fast gas chromatography/time-of-flight mass spectrometry for rapid characterization of synthetic polymers: Optimization of instrumental parameters, *J. Anal. Appl. Pyrol.*, 71(1), 2004, 327–341.

[39] P. F. Greenwood, S. C. George, M. A. Wilson, K. J. Hall, A new apparatus for laser micropyrolysis — gas chromatography/mass spectrometry, *J. Anal. Appl. Pyrol.*, 38, 1996, 101–118.

[40] T. F. da Silva, J. G. M. Filho, M. C. da Silva, A. D. de Oliveira, J. Torres de Souza, N. F. Rondon, *Botryococcus braunii versus Gloecapsomorpha prisca*: Chemical composition correlation using laser micropyrolysis-gas chromatography/mass spectrometer (LmPy-GCMSMS), *Int. J. Coal Geol.*, 168, 2016, 71–79.

[41] S. C. Moldoveanu (Ed.), *Techniques and Instrumentation in Analytical Chemistry, Vol. 28, Pyrolysis of Organic Molecules with Applications to Health and Environmental Issues*, Elsevier, Amsterdam, The Netherlands, 2010.

[42] N. A. al Sandouk-Lincke, *Application of Advanced Pyrolysis for the Analysis of Biogeochemically and Environmentally Significant Macromolecular Organic Materials: Microfossils, Macerals and Drilling Fluid Additives*, PhD Thesis, RWTH Aachen University, Aachen, Germany, 2014.

[43] J. Clayden, N. Greeves, S. Warren, P. Wothers, *Organic Chemistry*, 1st ed., Oxford University Press, Oxford, UK, 2000.

[44] E. R. Kaal, G. Alkemab, M. Kurano, M. Geissler, H.-G. Janssen, On-line size exclusion chromatography–pyrolysis–gas chromatography–mass spectrometry for copolymer characterization and additive analysis, *J. Chromatogr. A*, 1143, 2007, 182–189.

[45] R. Gächter, H. Müller, *Plastics Additives Handbook*, 4th ed., Hansa Publishers, Munich, Germany, 1993.

[46] M. Herrera, G. Matuschek, A. Kettrup, Fast identification of polymer additives by pyrolysis–gas chromatography/mass spectrometry, *J. Anal. Appl. Pyrol.*, 70, 2003, 35–42.

[47] P. Kusch, Identification of organic additives in nitrile rubber materials by pyrolysis-GC/MS, *LCGC North Am.*, 31(3), 2013, 248–254.

[48] B. M. Mandal, *Fundamentals of Polymerization*, World Scientific Publishing, Singapore, 2013.

[49] S. Tsuge, H. Ohtani, Microstructure of polyolefins, in *Applied Pyrolysis Handbook*, 2nd ed., T. Wampler (Ed.), CRC Press, Boca Raton, USA, 2007, p. 69.

[50] J. A. González-Pérez, N. T. Jiménez-Morillo, J. M. de la Rosa, G. Almendros, F. J. González-Vila, Pyrolysis-gas chromatography–isotope ratio mass spectrometry of polyethylene, *J. Chromatogr. A*, 1388, 2015, 236–243.

[51] K. V. Popov, V. D. Knyazev, Initial stages of the pyrolysis of polyethylene, *J. Phys. Chem. A*, 119, 2015, 11737–11760.

[52] L. M. Poutsma, Reexamination of the pyrolysis of polyethylene: Data needs, free-radical mechanistic considerations, and thermochemical kinetic simulation of initial product-forming pathways, *Macromol.*, 36, 2003, 8931–8957.

[53] I. J. Núñez Zorriqueta, *Pyrolysis of Polypropylene by Ziegler-Natta Catalysts*, Doctoral Thesis, Universität Hamburg, Hamburg, Germany, 2006.

[54] J. K. Y. Kiang, P. C. Uden, J. C. W. Chien, Polymer reactions. 7. Thermal pyrolysis of polypropylene, *Polym. Degrad. Stabil.*, 2(2), 1980, 113–127.

[55] T. M. W. Kruse, H.-W. Wong, L. J. Broadbelt, Mechanistic modelling of polymer pyrolysis: Polypropylene, *Macromol.*, 36, 2003, 9594–9607.

[56] S. Tsuge, H. Ohtani, Ch. Watanabe, *Pyrolysis-GC/MS Data Book of Synthetic Polymers; Pyrograms, Thermograms and MS of Pyrolyzates*, Elsevier, Amsterdam, The Netherlands, 2011.

[57] https://en.wikipedia.org/wiki/Polybutylene accessed Jan 3, 2017.

[58] https://www.lyondellbasell.com/en/products-technology/polymers/resin-type/polybutene-1/ accessed Jan 30, 2017.

[59] https://www.britannica.com/science/ethylene-propylene-copolymer accessed Aug 18, 2016.

[60] S.-S. Choi, Y.-K. Kim, Analysis of 5-ethylidene-2-norbornene in ethylene-propylene-diene terpolymer using pyrolysis-GC/MS, *Polym. Test.*, 30, 2011, 509–514.

[61] B. Andersen, *Investigations on Environmental Stress Cracking Resistance of LDPE/EVA Blends*, Doctoral Thesis, Martin-Luther-Universität Halle-Wittenberg, Halle (Saale), Germany, 2004.

[62] https://en.wikipedia.org/wiki/Ethylene-vinyl_acetate accessed Aug 24, 2016.

[63] K. M. Wiggins, Ch. W. Bielawski, Synthesis of poly(ethylene-*co*-acrylic acid) via a tandem hydrocarboxylation/hydrogenation of poly(butadiene), *Polym. Chem.*, 4, 2013, 2239–2245.

[64] http://www.dupont.com/products-and-services/plastics-polymers-resins/ethylene-copolymers/brands/nucrel-ethylene-acrylic-acid.html accessed Feb 13, 2017.

[65] P. Nordell, *Aluminium oxide — Poly(ethylene-co-butyl acrylate) Nanocomposites: Synthesis, Structure, Transport Properties and Long-term Performance*, Thesis, KTH Stockholm, Stockholm, Sweden, 2011.

[66] S. Nawaz, *Preparation and Long-term Performance of Poly(ethylene-co-butyl acrylate) Nanocomposites and Polyethylene*, Doctoral Thesis, KTH Stockholm, Stockholm, Sweden, 2012.

[67] DIN 13967 *Abdichtungsbahnen — Kunststoff- und Elastomerbahnen für die Bauwerksabdichtung gegen Bodenfeuchte und Wasser — Definitionen und Eigenschaften*, Beuth Verlag, Berlin, Germany, 2007.

[68] U. Lappan, U. Geißler, U. Scheler, Chemical structures formed in electron beam irradiated poly(tetrafluoroethylene-*co*-hexafluoropropylene) (FEP), *Macromol. Mater. Eng.*, 291, 2006, 937–943.

[69] P. Carlson, W. Schmiegel, Organic fluoropolymers, in *Ullman´s Encyclopedia of Industrial Chemistry*, B. Elvers (Ed.), Wiley-VCH, Weinheim, Germany, 2002.

[70] https://en.wikipedia.org/wiki/Polyvinyl_chloride accessed Feb 22, 2017.

[71] K. L. Sobeih, M. Baron, J. Gonzalez-Rodriguez, Recent trends and developments in pyrolysis–gas chromatography, *J. Chromatogr. A*, 1186, 2008, 51–66.

[72] M. M. Mlynek, *Quantitative Pyrolyse-Gaschromatographie-Massenspektrometrie-Kopplung zur Bestimmung der Zusammensetzung Vernetzter Polymere*, Doctoral Thesis, Technische Universität Darmstadt, Darmstadt, Germany, 2014 (in German).

[73] J. L. Bove, P. Dalven, Pyrolysis of phthalic acid esters: their fate, *Sci. Total Environ.*, 36(1), 1984, 313–318.

[74] K. Saido, T. Kuroki, T. Ikemura, M. Kirisawa, Studies on the thermal decomposition of phthalate esters. IX. Thermal decomposition of bis(2-ethylhexyl) phthalate, *J. Anal. Appl. Pyrol.*, 6, 1984, 171–181.

[75] S. K. Saxena, Chemical and Technical Assessment, Polyvinyl Alcohol, 61st JECFA, FAO, 2004.

[76] Y. Ma, T. Bai, F. Wang, The physical and chemical properties of the polyvinylalcohol/polyvinylpyrrolidone/hydroxyapatite composite hydrogel, *Mater. Sci. Eng. C*, 59, 2016, 948–957.

[77] V. V. Antić, M. P. Antić, A. Kronimus, K. Oing, J. Schwarzbauer, Quantitative determination of poly(vinylpyrrolidone) by continuous-flow off-line pyrolysis-GC/MS, *J. Anal. Appl. Pyrol.*, 90, 2011, 93–99.

[78] http://www.plasticseurope.org/what-is-plastic/types-of-plastics-11148/engineering-plastics/pvdf.aspx accessed Feb 28, 2017.

[79] N. Niessner, E. Jahnke, D. Wagner, M. Blinzler, C. Ruthard, Styrene copolymers, *Kunststoffe Int.*, 10, 2014, 28–33.

[80] C. Bouster, P. Vermande, J. Veron, Evolution of the product yield with temperature and molecular weight in the pyrolysis of polystyrene, *J. Anal. Appl. Pyrol.*, 15, 1989, 249–259.

[81] http://msdssearch.dow.com/PublishedLiteratureDOWCOM/dh_0988/0901b80380988639.pdf accessed Mar 8, 2017.

[82] https://www.britannica.com/science/rubber-chemical-compound accessed Mar 8, 2017.

[83] S.-S. Choi, Characteristics of the pyrolysis patterns of styrene-butadiene rubbers with differing microstructures, *J. Anal. Appl. Pyrol.*, 62, 2002, 319–330.

[84] T. Saito, Determination of styrene-butadiene and isoprene tire tread rubbers in piled particulate matter, *J. Anal. Appl. Pyrol.*, 15, 1989, 227–235.

[85] https://www.britannica.com/science/polyisoprene accessed Mar 13, 2017.

[86] http://iisrp.com/webpolymers/11polyisoprene.pdf accessed Mar 13, 2017.

[87] B. Danon, P. van der Gryp, C. E. Schwarz, J. F. Gorgens, A review of dipentene (dl-limonene) production from waste tire pyrolysis, *J. Anal Appl. Pyrol.*, 112, 2015, 1–13.

[88] http://pslc.ws/macrog/pib.htm accessed Apr 7, 2017.

[89] V. Dubey, S. K. Pandey, N. B. S. N. Rao, Research trends in the degradation of butyl rubber, *J. Anal. Appl. Pyrol.*, 34, 1995, 111–125.

[90] http://www.iisrp.com/webpolymers/01finalpolybutadienever2.pdf accessed Apr 7, 2017.

[91] D. F Graves, Rubber, in *Kent and Riegels Handbook of Industrial Chemistry and Biotechnology, Part 1*, J.A. Kent (Ed.), Springer Science + Business Media, New York, New York, USA, 2007, pp. 689–718.

[92] X. Liu, J. Zhao, Y. Liu, R. Yang, Volatile components changes during thermal aging of nitrile rubber by flash evaporation of Py-GC/MS, *J. Anal. Appl. Pyrol.*, 113, 2015, 193–201.

[93] M. Hakkarainen, Solid phase microextraction for analysis of polymer degradation products and additives, in *Advances in Polymer Science, Vol. 211, Chromatography for Sustainable Polymeric Materials*, A.-C. Albertsson and M. Hakkarainen (Eds.), Springer-Verlag, Berlin, Heidelberg, Germany, 2008, pp. 23–50.

[94] W. Buchberger, M. Stiftinger, Analysis of polymer additives and impurities by liquid chromatography/mass spectrometry and capillary electrophoresis/mass spectrometry, in *Advances in Polymer Science, Vol. 248, Mass Spectrometry of Polymers — New Techniques*, M. Hakkarainen (Ed.), Springer-Verlag, Berlin, Heidelberg, Germany, 2012, pp. 39–68.

[95] P. Kusch, V. Obst, D. Schroeder-Obst, W. Fink, G. Knupp, J. Steinhaus, Application of pyrolysis–gas chromatography/mass spectrometry for the identification of polymeric materials in failure analysis in the automotive industry, *Eng. Fail. Anal.*, 35, 2013, 114–124.

[96] P. Kusch, Application of pyrolysis–gas chromatography/mass spectrometry (Py-GC/MS), in *Comprehensive Analytical Chemistry*, Vol. 75, T. Rocha-Santos, A. Duarte (Eds.), Elsevier, Amsterdam, The Netherlands, 2017, pp. 169–207.

[97] F. Wang, Y. Liu, Z. Tang, M. Hou, Ch. Wang, X. Wang, Q. Wang, Q. Xiao, Simultaneous determination of 15 phthalate esters in commercial beverages using dispersive liquid–liquid microextraction coupled to gas chromatography-mass spectrometry, *Anal. Meth.*, 9, 2017, 1912–1919.

[98] http://www.iisrp.com/webpolymers/04finalpolychloropreneiisrp.pdf accessed Apr 11, 2017.

[99] http:/polymerdatabase.com/Elastomers/Chloroprene.html accessed Apr 11, 2017.

[100] A. J. Tinker, Introduction — the book and rubber blends, in *Blends of Natural Rubber, Novel Techniques for Blending with Specialty Polymers*, A. J. Tinker, K. P. Jones (Eds.), Chapman & Hall, London, Springer Netherlands, 1998, pp. 1–7.

[101] U. Šebenik, A. Zupančič-Valant, M. Krajnc, Investigation of rubber–rubber blends miscibility, *Polym. Eng. Sci.*, 46(11), 2006, 1849–1659.

[102] P. Kusch, Application of pyrolysis-gas chromatography/mass spectrometry in failure analysis in the automotive industry, in *Automobiles and the Automotive Industry: Emerging Technologies, Environmental Impact and Safety Analysis*, A. T. Evans (Ed.), Nova Science Publishers, New York, USA, 2015, pp. 41–64.

[103] A. Rathi, M. Hernández, W. K. Dierkes, J. W. M. Noordermeer, C. Bergmann, J. Trimbach, A. Blume, Effect on aromatic oil on phase dynamics of S-SBR/BR blends for passenger car tire treads. Paper presented at the Fall 190th Technical Meeting of the Rubber Division of the American Chemical Society, Inc., Pittsburgh, PA, USA, October 12, 2016, ISSN: 1547-1977.

[104] S. C. George, K. N. Ninan, G. Groeninckx, S. Thomas, Styrene–butadiene rubber/natural rubber blends: Morphology, transport behavior, and dynamic mechanical and mechanical properties, *J. Appl. Polym. Sci.*, 78, 2000, 1280–1303.

[105] J. T. Varkey, S. Augustine, S. Thomas, Thermal degradation of natural rubber/styrene butadiene rubber latex blends by thermogravimetric method, *Polym. Plast. Technol. Eng.*, 39(3), 2000, 415–435.

[106] http://www.seamate.com.tw/csr.html accessed Sep 8, 2017.

[107] W. G. Fischer, P. Kusch, Quantitative Bestimmung von Polymermischungen, *CLB Chemie in Labor und Biotechnik*, 47(1), 1996, 4–7 (in German).

[108] E. Andrzejewska, P. Kusch, A. Andrzejewski, Thermal decomposition of crosslinked polymers of some dimethylacrylate esters, *Polym. Degrad. Stab.*, 40(1), 1993, 27–30.

[109] M. Kamberi, D. Pinson, S. Pacetti, L. E. L. Perkins, S. Hossainy, H. Mori, R. J. Rapoza, F. Kolodgie, R. Virmani, Evaluation of chemical stability of polymers of XIENCE everolimus-eluting coronary stents *in vivo* by pyrolysis-gas chromatography/mass spectrometry, *J. Biomed. Mater. Res., Part B Appl. Biomater.*, 106B(5), 2018, 1721–1729.

[110] R. Rogalewicz, A. Voelkel, I. Kownacki, Application of HS-SPME in the determination of potentially toxic organic compounds emitted from resin-based dental materials, *J. Environ. Monit.*, 8(3), 2006, 377–383.

[111] R. Rogalewicz, K. Batko, A. Voelkel, Identification of organic extractables from commercial resin-modified glass-ionomers using HPLC-MS, *J. Environ. Monit.*, 8(7), 2006, 750–758.

[112] P. Kusch, C. Rieser, G. Knupp, T. Mang, Characterization of copolymers of methacrylic acid with poly(ethylene glycol) methyl ether methacrylate macromonomers by analytical pyrolysis-gas chromatography/mass spectrometry (Py-GC/MS), *J. Anal. Appl. Pyrol.*, 113, 2015, 412–418.

[113] A. Bernnat, P. Eibeck, J. Dirlenbach, Strong in electronics and automotive construction: Polybutylene terephthalate (PBT), *Kunststoffe*, 10, 2013, 106–110.

[114] E. J. Dziwiński, J. Iłowska, J. Gniady, Py-GC/MS analyses of poly(ethylene terephthalate) film without and with the presence of tetramethylammonium acetate reagent. Comparative study, *Polym. Test.*, 65, 2018, 111–115.

[115] K. Kawai, H. Kondo, H. Ohtani, Characterization of cross-linking structure in terephthalate polyesters formed through material recycling process by pyrolysis-gas chromatography in the presence of organic alkali, *Polym. Degrad. Stab.*, 93, 2008, 1781–1785.

[116] D. J. Brunelle, Polycarbonates, in *Encyclopedia of Polymer Science and Technology*, Wiley Online Library, 2006. DOI: 10.1002/0471440264.pst255.pub2.

[117] https://www.creativemechanisms.com/blog/everything-you-need-to-know-about-polycarbonate-pc accessed Oct 27, 2017.

[118] M. Day, J. D. Cooney, D. M. Wiles, The thermal degradation of poly(aryl-ether-ether-ketone) (PEEK) as monitored by pyrolysis-GC/MS and TG/MS, *J. Anal. Appl. Pyrol.*, 18(2), 1990, 163–173.

[119] C. J. Tsai, L. H. Perng, Y. C. Ling, A study of thermal degradation of poly(aryl-ether-ether-ketone) using stepwise pyrolysis/gas chromatography/ mass spectrometry, *Rapid Comm. Mass Spectrom.*, 11, 1997, 1987–1995.

[120] L. H. Perng, C. J. Tsai, Y. C. Ling, Mechanism and kinetic modelling of PEEK pyrolysis by TG/MS, *Polymer*, 40(26), 1999, 7321–7329.

[121] https://en.wikipedia.org/wiki/Polyoxymethylene accessed Nov 7, 2017.

[122] W. F. Gum, W. Riese, H. Ulrich (Eds.), *Reaction Polymers*, Hanser Publisher, München, Germany, 1992.

[123] https://de.wikipedia.org/wiki/Bakelite accessed Dec 4, 2017.

[124] H. Jiang, J. Wang, S. Wu, Z. Yuan, Z. Huc, R. Wu, Q. Liu, The pyrolysis mechanism of phenol formaldehyde resin, *Polym. Degrad. Stab.*, 97, 2012, 1527–1533.

[125] Phenol-resorcinol-formaldehyde resins and the process of making them, using an alkali metal sulfite catalyst, Patent US 3328354 A, 1967.

[126] V. M. Gun'ko, V. M. Bogatyrov, O. I. Oranska, I. V. Urubkov, R. Leboda, B. Charmas, J. Skubiszewska-Zięba, Synthesis and characterization of resorcinol–formaldehyde resin chars doped by zinc oxide, *Appl. Surf. Sci.*, 303, 2014, 263–271.

[127] N. W. Schwandt, T. G. Gound, Resorcinol-formaldehyde resin "Russian red" endodontic therapy, *J. Endod.*, 29(7), 2003, 435–437.

[128] http://polymerdatabase.com/polymer%20classes/MelamineFormaldehyde %20type.html accessed Dec 4, 2017.

[129] E. Sharmin, F. Zafar, Polyurethane: An introduction, in *Polyurethane*, F. Zafar (Ed.), InTech, Rijeka, Croatia, 2012, pp. 3–16. DOI: 10.5772/51663. https://www.intechopen.com/books/polyurethane/polyurethane-an-introduction accessed Dec 4, 2017.

[130] P. Kusch, Application of headspace-solid phase microextraction (HS-SPME) coupled with gas chromatography/mass spectrometry (GC/MS) for the characterization of polymers, in *Gas Chromatography: Analysis, Methods and Practices*, V. Waren (Ed.), Nova Science Publishers Inc., New York, USA, 2017, pp. 69–104.

[131] J. A. Hiltz, Analytical pyrolysis gas chromatography/mass spectrometry (Py-GC/MS) of poly(ether urethane)s, poly(ether urea)s and poly(ether urethane-urea)s, *J. Anal. Appl. Pyrol.*, 113, 2015, 248–258.

[132] https://en.wikipedia.org/wiki/Aramid accessed Dec 8, 2017.

[133] P. Kusch, G. Knupp, W. Fink, D. Schroeder-Obst, V. Obst, J. Steinhaus, Application of pyrolysis–gas chromatography–mass spectrometry for the identification of polymeric materials, *LCGC North Am.*, 32(3), 2014, 210–217.

[134] A. S. Rahate, K. R. Nemade, S. A. Waghuley, Polyphenylene sulfide (PPS): state of the art and applications, *Rev. Chem. Eng.*, 29(6), 2013, 471–489.

[135] L. A. Goettler, J. J. Scobbo, Applications of polymer blends, in *Polymer Blends Handbook*, L. A. Utracki, C. A. Wilkie (Eds.), Springer Science+ Business Media BV, Dordrecht, The Netherlands, 2014, pp. 1433–1458.

[136] S. Alarcon Salinas, P. Kusch, G. Knupp, J. Steinhaus, D. Sülthaus, Characterization and quantification of poly(acrylonitrile-*co*-1,3-butadiene-*co*-styrene)/polyamide 6 (ABS/PA6) blends using pyrolysis-gas chromatography (Py-GC) with different detector systems, *J. Anal. Appl. Pyrol.*, 122, 2016, 452–457.

[137] L. M. Robeson, Applications of polymer blends: Emphasis on recent advances, *Polym. Eng. Sci.*, 24(8), 1984, 587–597.

[138] M. K. Akkapeddi, Commercial polymer blends, in *Polymer blends handbook*, L. A. Utracki (Ed.), Kluwer Academic Publishers, Dordrecht, The Netherlands, 2002, p. 1039.

[139] R. A. Kudva, H. Keskkula, D. R. Paul, Properties of compatibilized nylon 6/ ABS blends: Part I. Effect of ABS type, *Polymer*, 41, 2000, 225–237.

[140] W. Arayapranee, P. Prasassarakich, G. L. Rempel, Blends of poly(vinyl chloride) (PVC)/natural rubber-*g*-(styrene-*co*-methyl methacrylate) for improved impact resistance of PVC, *J. Appl. Polym. Sci.*, 93(4), 2004, 1666–1672.

[141] M. S. A. Moraes, M. V. Migliorini, F. C. Damasceno, F. Georges, S. Almeida, C. A. Zini, R. A. Jacques, E. B. Caramão, Qualitative analysis of bio oils of agricultural residues obtained through pyrolysis using comprehensive two-dimensional gas chromatography with time-of-flight mass spectrometric detector, *J. Anal. Appl. Pyrol.*, 98, 2012, 51–64.

[142] S. Wu, G. Lv, R. Lou, Applications of chromatography hyphenated techniques in the field of lignin pyrolysis, in *Applications of Gas Chromatography*, R. Davarnejad (Ed.), InTech, Rijeka, Croatia, 2012, pp. 41–64. http://www.intechopen.com/books/applications-of-gaschromatography/applications-of-chromatography-hyphenated-techniques-in-the-field-of-lignin-pyrolysis accessed Feb 14, 2018.

[143] H. Cheng, S. Wu, X. Li, Comparison of the oxidative pyrolysis behaviors of black liquor solids, alkali lignin and enzymatic hydrolysis/mild acidolysis lignin, *RSC Adv.*, 5, 2015, 79532–79537.

[144] B. Hansen, P. Kusch, M. Schulze, B. Kamm, Qualitative and quantitative analysis of lignin produced from beech wood by different conditions of the organosolv process, *J. Polym. Environ.*, 24, 2016, 85–97.

[145] M. Zhang, F. L. P. Resende, A. Moutsoglou, D. E. Raynie, Pyrolysis of lignin extracted from prairie cordgrass, aspen, and Kraft lignin by Py-GC/MS and TGA/FTIR, *J. Anal Appl. Pyrol.*, 98, 2012, 65–71.

[146] M. Sobiesiak, B. Podkościelna, O. Sevastyanova, Thermal degradation behavior of lignin-modified porous styrene-divinylbenzene and styrene-bisphenol A glycerolate diacrylate copolymer microspheres, *J. Anal. Appl. Pyrol.*, 123, 2017, 364–375.

[147] Y.-L. Hsieh, Chemical structure and properties of cotton, in *Cotton: Science and technology*, S. Gordon, Y.-L. Hsieh (Eds.), CRC Press, Boca Raton, USA, 2007, pp. 3–30.

[148] X. Zhang, W. Yang, Ch. Dong, Levoglucosan formation mechanisms during cellulose pyrolysis, *J. Anal. Appl. Pyrol.*, 104, 2013, 19–27.

[149] G. C. Galletti, P. Bocchini, Pyrolysis/gas chromatography/mass spectrometry of lignocellulose, *Rapid Commun. Mass Spectrom.*, 9, 1995, 815–826.

[150] https://en.wikipedia.org/wiki/Chitosan accessed Feb 14, 2018.

[151] W. G. Root, J. van Krieken, O. Sliekesl, S. de Vos, Production and purification of lactic acid and lactide, in *Poly(lactic acid): Synthesis, Structures, Properties, Processing, and Applications*, R. A. Auras, L.-T. Lim, S. E. M. Selke, H. Tsuji (Eds.), Wiley, Hoboken, New Jersey, USA, 2010, pp. 3–18.

[152] F. Shadkami, R. Helleur, Recent applications in analytical thermochemolysis, *J. Anal. Appl. Pyrol.*, 89(1), 2010, 2–16.

[153] A. Venema, R. C. A. Boom-van Geest, In-situ hydrolysis/methylation pyrolysis GC for the characterization of polyaramides, *J. Microcolumn Sep.*, 7(4), 1995, 337–343.

[154] Y. Ishida, H. Ohtani, K. Abe, S. Tsuge, K. Yamamoto, K. Katoh, Sequence distributions of polyacetals studied by reactive pyrolysis-gas chromatography in the presence of cobalt sulfate, *Macromol.*, 28, 1995, 6528–6532.

[155] J. M. Challinor, Characterisation of wood by pyrolysis derivatization — gas chromatography/mass spectrometry, *J. Anal. Appl. Pyrol.*, 35, 1995, 93–107.

[156] Y. Ishida, K. Ohsugi, K. Taniguhi, H. Ohtani, Rapid determination of terephthalic acid in the hydrothermal decomposition product of poly(ethylene terephthalate) by thermochemolysis–gas chromatography in the presence of tetramethylammonium acetate, *Anal. Sci.*, 27, 2011, 1053–1056.

[157] H. Ohtani, S. Tsuge, Characterization of condensation polymers by pyrolysis-GC in the presence of organic alkali, in *Applied Pyrolysis*

Handbook, 2nd ed., T. P. Wampler (Ed.), CRC Press, Boca Raton, FL, USA, 2007, pp. 249–269.

[158] D. Tamburini, I. Bonaduce, M. P. Colombini, Characterisation of oriental lacquers from *Rhus succedanea* and *Melanorrhoea usitata* using *in situ* pyrolysis/silylation-gas chromatography-mass spectrometry, *J. Anal. Appl. Pyrol.*, 116, 2015, 129–141.

[159] L. Osete-Cortina, M. T. Doménech-Carbó, Characterization of acrylic resins used for restoration of artworks by pyrolysis-silylation-gas chromatography/mass spectrometry with hexamethyldisilazane, *J. Chromatogr. A*, 1127, 2006, 228–236.

[160] J. La Nasa, M. Zanaboni, D. Uldanck, I. Degano, F. Modugno, H. Kutzke, E. Storevik Tveit, B. Topalova-Casadiego, M. P. Colombini, Novel application of liquid chromatography/mass spectrometry for the characterization of drying oils in art: Elucidation on the composition of original paint materials used by Edvard Munch (1863–1944), *Anal. Chim. Acta*, 896, 2015, 177–189.

[161] K. MacNamara, R. Leardi, A. Hoffmann, Developments in 2-D gas chromatography with heartcutting, *LCGC North Am.*, 22(2), 2004, 166–186.

[162] D. Deans, A new technique for heart cutting in gas chromatography, *Chromatographia*, 1, 1968, 18–22.

[163] W. J. Bertsch, Methods in high resolution gas chromatography: Two-dimensional techniques, *J. High Resolut. Chromatogr. Chromatogr. Commun.*, 1(2), 1978, 85–90.

[164] R. E. Kaiser, R. I. Rieder, L. Leming, L. Blomber, P. Kusz, Dramatic selectivity changes on carrier gas flow changes in a series-coupled GC capillary tandem, *J. High Resolut. Chromatogr. Chromatogr. Commun.*, 8(9), 1985, 580–584.

[165] Z. Liu, J. B. Phillips, Comprehensive two-dimensional gas chromatography using an on-column thermal modulator interface, *J. Chromatogr. Sci.*, 29, 1991, 227–231.

[166] Comprehensive GC system based on flow modulation for the 7890A GC, Application Brief, Agilent Technologies, Wilmington, DE, USA, 2008.

[167] J. V. Hinshaw, Comprehensive-two-dimensional gas chromatography, *LCGC Eur.*, 17(2), 2004, 86–95.

[168] F. Cheng-Yu Wang, C. C. Walters, Pyrolysis comprehensive two-dimensional gas chromatography: Study of petroleum source rock, *Anal. Chem.*, 79, 2007, 5642–5650.

[169] S. Scherer, Comprehensive two-dimensional gas chromatography coupled with time-of-flight mass spectrometry for broad spectrum organic analysis (GC×GC-TOFMS), University of Michigan, Ann Arbor, USA, 2005. http://

www.hems-workshop.org/5thWS/Talks/Thursday/SCHERER.pdf accessed Feb 22, 2018.

[170] Chemical analysis of polymer microbeads in toothpaste by TD- and Py-GC×GC-TOFMS, Application Note, LECO Corporation, Saint Joseph, Michigan, USA, 2016.

[171] B. Hana, J. Vialb, M. Inabac, M. Sablier, Analytical characterization of East Asian handmade papers: A combined approach using Py-GC×GC/MS and multivariate analysis, *J. Anal. Appl. Pyrol.*, 127, 2017, 150–158.

[172] R. M. Twyman, P. Kusch, Mass spectrometry: Pyrolysis, in *Reference Module in Chemistry, Molecular Sciences and Chemical Engineering*, J. Reedijk (Ed.), Elsevier, Amsterdam, The Netherlands, 2017.

[173] A. Shiono, A. Hosaka, C. Watanabe, N. Teramae, N. Nemoto, H. Ohtani, Thermoanalytical characterization of polymers: A comparative study between thermogravimetry and evolved gas analysis using a temperature-programmable pyrolyzer, *Polym. Test.*, 42, 2015, 54–61.

[174] http://www.plasticseurope.org/documents/document/20151216062602-plastics_the_facts_2015_final_30pages_14122015.pdf accessed Aug 18, 2016.

[175] http://www.analytik-news.de/Presse/2015/138.html accessed Aug 11, 2016.

[176] http://www.oceanconservancy.org/our-work/marine-debris/iccdata-2014.pdf accessed Aug 11, 2016.

[177] G. Duessing, A sea of polymers, GERSTEL Solutions worldwide, No. 16, 2015.

[178] M. Cole, P. Lindeque, C. Halsband, T. S. Galloway, Microplastics as contaminants in the marine environment: A review, *Marine Poll. Bull.*, 62, 2011, 2588–2597.

[179] https://www.bmbf.de/pub/The_Future_of_the_Oceans.pdf accessed Aug 11, 2016.

[180] European Parliament and the Council, Directive 2008/56/EC of 17 June 2008 establishing a framework for community action in the field of marine environmental policy (Marine Strategy Framework Directive), *Off. J. Eur. Union*, 51, 2008, L164, 19–40.

[181] European Commission, Commission Decision 2010/477/EU of 1 September 2010 on criteria and methodological standards on good environmental status of marine waters (notified under document C 2010 5956), *Off. J. Eur. Union*, 53, 2010, L232, 14–24.

[182] E. Fries, J. H. Dekiff, J. Willmeyer, M.-T. Nuelle, M. Ebert, D. Remy, Identification of polymer types and additives in marine microplastic particles using pyrolysis-GC/MS and scanning electron microscopy, *Environ. Sci. Process. Impacts*, 15, 2013, 1949–1956.

[183] JRC, MSFD GES Technical Subgroup on Marine Litter. *Marine Litter — Technical Recommendations for the Implementation of MSFD Requirements*, Luxembourg, 2011.

[184] T. Rocha-Santos, A. C. Duarte, A critical overview of the analytical approaches to the occurrence, the fate and the behavior of microplastics in the environment, *Trends Anal. Chem.*, 65, 2015, 47–53.

[185] V. Hidalgo-Ruz, L. Gutow, R. C. Thompson, M. Thiel, Microplastics in the marine environment: A review of the methods used for identification and quantification, *Environ. Sci. Technol.*, 46, 2012, 3060–3075.

[186] M. G. J. Löder, G. Gerdts, Methodology used for the detection and identification of microplastics — a critical appraisal, in *Marine Anthropogenic Litter*, M. Bergmann, L. Gutow, M. Klages (Eds.), Springer, Cham, Heidelberg, Germany, 2015, pp. 201–227.

[187] P. Kusch, Analyzing failure using pyrolysis-GC-MS, *The Column*, 10(19), 2014, 17–20.

[188] W. Gao, M. Chen, Characterizing the average composition and molar mass distributions of a poly(styrene-*co*-acrylic acid) copolymer by SEC-MALS-dRI-UV, Application Note, Wyatt Technology Co., Santa Barbara, CA, USA, 2018.

[189] http://www.chromatographyonline.com/synthetic-rubbers-polybutadiene-1 accessed Feb 28, 2018.

[190] P. Kusch, Identification of synthetic polymers and copolymers by analytical pyrolysis — gas chromatography/mass spectrometry, *J. Chem. Educ.*, 91(10), 2014, 1725–1728.

Appendix A

Pure & Appl. Chem., Vol. 65, No. 11, pp. 2405–2409, 1993.
Printed in Great Britain

INTERNATIONAL UNION OF PURE AND APPLIED CHEMISTRY

ANALYTICAL CHEMISTRY DIVISION COMMISSION ON CHROMATOGRAPHY AND OTHER

ANALYTICAL SEPARATIONS*

NOMENCLATURE AND TERMINOLOGY FOR

ANALYTICAL PYROLYSIS (IUPAC Recommendations 1993)

PETER C. UDEN

Department of Chemistry, University of Massachusetts, Amherst, Massachusetts, 01003, USA

*Members of the Commission during the period (1989–1993) when this report was prepared was as follows: *Chairman:* P. C. Uden (USA, 1989–1993); *Secretary:* C. A. M. G. Cramers (Netherlands, 1989–1991); R. M. Smith (UK, 1991–1993); *Titular Members:* H. M. Kingston (USA, 1989–1993); A. Marton (Hungary, 1991–1993); *Associate Members:*

V. A. Davankov (USSR, 1991–1993); F. M. Everaerts (Netherlands, 1989–1993); K. Jinno (Japan, 1991–1993); J. A. Jonsson (Sweden, 1991–1993); A. Marton (Hungary, 1989–1991); R. M. Smith (UK, 1989–1991); G. Vigh (1989–1991); W. Yu (China, 1989–1993); *National Representatives:* R. M. Habib (Egypt, 1990–1993); F. Radler de Aquino Neto (1991–1993); J. Garaj (Czechoslovakia, 1989–1991); P. Boček (Czechoslovakia, 1991–1993); D. Baylocq (France, 1989–1993); W. Engelwald (Germany, 1989–1993); P. A. Siskos (Greece, 1989–1993); S. N. Tandon (India, 1989–1993); D. W. Lee (Korea, 1991–1993); J. A. Garcia Dominguez (Spain, 1991–1993); S. Ozden (Turkey, 1991–1993); U. L. Haldna (USSR/Estonia).

Names of countries given after Members' names are in accordance with the *IUPAC Handbook 1991–1993*; changes will be effected in the 1993–1995 edition.

Nomenclature and Terminology for Analytical Pyrolysis (IUPAC Recommendations 1993)

Abstract

This paper defines terms and definitions used in analytical methods of pyrolysis and includes expressions for coupled systems and for the description of the temperature profiles and the products that are obtained.

Introduction

Thermal degradation under controlled conditions is often used an part of an analytical procedure, either to render a sample into a suitable form for subsequent analysis by gas chromatography, mass spectrometry, or infrared spectroscopy, or by direct monitoring as analytical technique in its own right. A range of terms and expression have been used in the field and this nomenclature brings these together in a systematic manner and assigns each a specific meaning.

Analytical Pyrolysis

***Analytical pyrolysis*:** The characterization, in an inert atmosphere, of a material or a chemical process by a chemical degradation reaction(s) induced by thermal energy.

***Catalytic pyrolysis*:** A pyrolysis that is influenced by the addition of a catalyst.

***Char*:** A solid carbonaceous pyrolysis residue.

***Coil pyrolyser*:** A pyrolyser in which the sample (sometimes located in a tubular vessel) is placed in a metal coil that is heated to cause pyrolysis.

***Continuous mode (furnace) pyrolyser*:** A pyrolyser in which the sample is introduced into a furnace preheated to the final temperature.

***Curie-point pyrolyser*:** A pyrolyser in which a ferromagnetic sample carrier is inductively heated to its Curie-point.

***Filament (ribbon) pyrolyser*:** A pyrolyser in which the sample is placed on a metal filament (ribbon) that is resistively heated to cause pyrolysis.

Final pyrolysis temperature **($T_{(f,\ Py)}$):** The final (steady state) temperature which is attained by a pyrolyser. (The terms "equilibrium temperature" and "pyrolysis temperature" may be used when referring to an isothermal pyrolysis; they are not recommended for use with a nonisothermal pyrolysis).

***Flash pyrolysis*:** A pyrolysis that is carried out with a fast rate of temperature increase, of the order of 10,000 K/s.

***Fractionated pyrolysis*:** A pyrolysis in which the same sample is pyrolyzed at different temperatures for different times in order to study special fractions of the sample.

***In-source pyrolysis*:** A pyrolysis in which the reactor is located within the ion source of a mass spectrometer.

***IR-pyrogram*:** Chromatogram of a pyrolysate detected by infrared spectrometry.

***Isothermal pyrolysis*:** A pyrolysis during which the temperature is essentially constant.

Maximum pyrolysis temperature **($T_{(max, Py)}$):** The highest temperature in a temperature/time profile.

***MS-pyrogram*:** Chromatogram of a pyrolysate detected by mass spectrometry.

***Off-line pyrolysis*:** A pyrolysis in which the products are trapped before analysis.

***Oxidative pyrolysis*:** A pyrolysis that occurs in the presence of an oxidative atmosphere.

***Pressure monitored pyrolysis*:** A pyrolysis technique in which the pressure of the volatile pyrolysates is recorded as the sample is heated.

***Pulse mode pyrolyser*:** A pyrolyser in which the sample is introduced into a cold furnace which is then heated rapidly.

***Pyrogram*:** A chromatogram of a pyrolysate.

Pyrolysate **(*pyrolyzate*):** The products of pyrolysis.

Pyrolyser **(*pyrolyzer*):** A device for performing pyrolysis.

Pyrolysis*:** A chemical degradation reaction that is caused by thermal energy. (The term ***pyrolysis generally refers to an inert environment.)

Pyrolysis–gas chromatography **(*Py–GC*):** A pyrolysis technique in which the volatile pyrolysates are directly conducted into a gas chromatograph for separation and detection.

Pyrolysis–gas chromatography–mass spectrometry **(*Py–GC–MS*):** A pyrolysis technique in which the volatile pyrolysates are separated and analyzed by online gas chromatography–mass spectrometry.

Pyrolysis–gas chromatography–infrared spectroscopy **(*Py–GC–IR*):** A pyrolysis technique in which the volatile pyrolysates are separated and analyzed by online gas chromatography–infrared spectroscopy.

Pyrolysis–infrared spectroscopy **(*Py–IR*):** A pyrolysis technique in which the pyrolysates are detected and analyzed by online infrared spectroscopy.

***Pyrolysis–infrared spectrum*:** Infrared spectrum obtained from pyrolysis-infrared spectroscopy.

***Pyrolysis–mass spectrometry* (*Py–MS*):** A pyrolysis technique in which the volatile pyrolysates are detected and analyzed by online mass spectrometry.

***Pyrolysis–mass spectrum*:** Mass spectrum obtained from *pyrolysis–mass spectrometry.*

***Pyrolysis reactor*:** The portion of the pyrolyser in which the pyrolysis takes place.

***Pyrolysis residue*:** The portion of the pyrolysate that does not leave the reactor.

***Pyrolysis thermogram*:** The result of a temperature-programmed pyrolysis in which the detector signals, e.g. total ion current or single ions, total absorbance or a GC detector, are plotted against time or temperature.

***Reductive pyrolysis*:** A pyrolysis which occurs in the presence of a reducing atmosphere.

***Sequential pyrolysis*:** A pyrolysis in which the same initial sample is repetitively pyrolyzed under identical conditions (final pyrolysis temperature, temperature-rise time and total heating time).

***Stepwise pyrolysis*:** A pyrolysis in which the sample temperature is raised stepwise. The pyrolysis products are recorded between each step.

***Tar*:** A liquid pyrolysis residue.

***Temperature-programmed pyrolysis*:** A pyrolysis during which the sample is heated at a controlled rate within a temperature range in which pyrolysis occurs.

***Temperature-rise time* (*TRT*):** The time required for a pyrolyser temperature to be increased from its initial to its final temperature.

***Temperature–time profile* (*TTP*):** A graphical representation of temperature *versus* time for a particular pyrolysis experiment or pyrolyser.

Total heating time (THT): The time between the onset and conclusion of the sample heating in a pyrolysis experiment.

Volatile pyrolyzate: The portion of the pyrolysate which has adequate vapor pressure to reach the detector.

List of Symbols

$T_{(f, Py)}$	Final pyrolysis temperature
$T_{(max, Py)}$	Maximum pyrolysis temperature

Index of Acronyms

Py–GC	Pyrolysis–gas chromatography
Py–GC–IR	Pyrolysis–gas chromatography–infrared spectroscopy
Py–GC–MS	Pyrolysis–gas chromatography–mass spectrometry
Py–IR	Pyrolysis–infrared spectroscopy
Py–MS	Pyrolysis–mass spectrometry
THT	Total heating time
TRT	Temperature-rise time
TTP	Temperature–time profile

Appendix B

Characteristic pyrolysis products of the most widely used polymers and copolymers at 700°C [96, 190].

Polymer/copolymer	CAS registry number	Pyrolysis products
PE	9002-88-4	α,ω-Alkadienes, α-alkenes, n-alkanes
PP	9003-07-0	2-Methyl-1-pentene, 2,4-dimethyl-1-heptene, 2,4,6-trimethyl-1-nonene, 2,4,6,8-tetramethyl-1-undecene
Poly(ethylene-*co*-vinyl acetate) (EVA)	24937-78-8	α, ω-Alkadienes, α-alkenes, n-alkanes, acetic acid
PMMA	9011-14-7	MMA, MA
PBA	9003-49-0	n-Butyl acrylate, n-butanol, n-butane
PS	9003-53-6	Styrene, benzene, toluene, α-methylstyrene, stilbene, bibenzyl, 1,2-diphenylethylene, styrene dimer
Polybutadiene (BR)	9003-17-2	1,3-Butadiene, benzene, toluene, 4-vinylocyclo-1-hexene
PVC	9002-86-2	Hydrogen chloride, benzene, toluene, styrene, naphthalene
PTFE (Teflon)	9002-84-0	TFE

(*Continued*)

(*Continued*)

Polymer/copolymer	CAS registry number	Pyrolysis products
Polyisoprene (NR)	9003-31-0	2-Methyl-1,3-butadiene (isoprene), dipentene (isoprene dimer)
Poly(styrene-*co*-1,3-butadiene) rubber (SBR)	9003-55-8	1,3-Butadiene, benzene, toluene, 4-vinylocyclo-1-hexene, ethylbenzene, styrene, α-methylstyrene
Poly(acrylonitrile-*co*-1,3-butadiene) (NBR)	9003-18-3	1,3-Butadiene, acrylonitrile, 1,3-cyclopentadiene, methacrylonitrile, benzene, toluene, styrene, benzonitrile
Poly(2-chloro-1,3-butadiene) (CR, chloroprene rubber)	9010-98-4	Hydrogen chloride, 2-chloro-1,3-butadiene, 1-chloro-5-(1-chloroethenyl)-cyclohexene, 1-chloro-4-(1-chloroethenyl)-cyclohexene
SAN	9003-54-7	Styrene, acrylonitrile
ABS	9003-56-9	Acrylonitrile, methacrylonitrile, 1,3-butadiene, styrene, α-methylstyrene, toluene, ethylbenzene, benzene-propanenitrile, benzenebutanenitrile, 1,2-diphenylethane
ABS-α-MS	162680-79-7	Acrylonitrile, 1,3-butadiene, benzene, toluene, 4-vinylocyclo-1-hexene, styrene, α-methylstyrene
PF resin	9003-35-4	Benzene, phenol, cresols, xylenols
PC	25766-59-0	Phenol, cresols, *p*-ethylphenol, 2-(4-hydroxyphenyl)-propane, 2-(4-hydroxyphenyl)-propene, 1-methyl-4-isopropenylphenol, 2,2-(4,4′-dihydroxydiphenyl)-propane (bisphenol A)
Polycaprolactam (PA 6, Nylon 6)	25038-54-4	Caprolactam, mononitriles
Poly(hexamethylene adipamide) (PA 6-6, Nylon 6-6)	32131-17-2	Cyclopentanone, capronitrile, hexanedinitrile
Poly[imino(1-oxo-1,12-dodecanediyl)] (PA 12, Nylon12)	24937-16-4	α-Alkenes, *n*-alkanes, alkylnitriles

(*Continued*)

(*Continued*)

Polymer/copolymer	CAS registry number	Pyrolysis products
Polyurethane (PU) (1,4-Butanediol-1,3-diisocyanatomethyl-benzene polymer)	37338-53-7	C_3–C_6 hydrocarbons, benzene, aniline, isocyanatotoluene, diisocyanatotoluene
PET	25038-59-9	Benzene, acetophenone, benzoic acid, terephthalic acid, biphenyl, *p*-diacetylbenzene
PEEK	31694-16-3	Phenol, *p*-benzoquinone, diphenylether, dibenzofuran, *p*-phenoxyphenol, 1,4-diphenoxybenzene
PDMS	9016-00-6	Cyclic oligodimethylsiloxanes
Poly(oxymethylene) (POM)	9015-98-9	Formaldehyde, methanol
PVA	9002-89-5	Acetaldehyde, methylbutene, acetic acid, 2-butenal
Poly(*N*-vinyl-2-pyrrolidone) (PVP)	9003-39-8	*N*-Vinyl-2-pyrrolidone, 1-methyl-2-pyrrolidone, 2-pyrrolidone

Index

Printed in the United States
By Bookmasters